A Window On Whaling
In British Columbia

Joan Goddard

Canadian Cataloguing in Publication Data

Goddard, Joan.
   A window on whaling in British Columbia

Includes index.
ISBN 1-895332-14-1

1. Whaling—British Columbia—History. I. Title.
SH383.5.C3G62 1997   338.3'7295'09711   C97-910834-9

Design & production by:
   Desktop Publishing Ltd.   Phone: (250) 383-8622
   2604 Quadra St.   Email: desktop@islandnet.com
   Victoria, BC V8T 4E4   Web: www.islandnet.com/~desktop

Cover painting by:
   Harry Heine, R.S.M.A., F.C.A.

Printed in Canada by:
   Premier Printing Ltd.
   Winnipeg, Manitoba

Contents

DEDICATION

This book is dedicated to the memory of my grandfather, William Rolls, who, with other Newfoundland men, came west in the first decade of this century to supervise the new whaling industry. To a small child, his dwelling place "up north at the station" was as incomprehensible as God's was in Heaven. He died at Naden Harbour in February 1933.

ACKNOWLEDGEMENTS

This book began with conversation...lots of it with people I had seen in my grandparents' photo album and hunted up. Lasting friendships have been the greatest legacy of this project. Heartfelt thanks go to all my informants in Canada, the United States and in Norway. The late Alan Armour, himself a veteran of the whaling industry in BC, was a tireless researcher who shared most generously.

Many thanks go to the following people who assisted in the production of this book: Merle Armour, Graeme Balcom, Sharon Cordova, Jim Darling, Joan Druett, Graeme Ellis, Richard Ellis, Milton Freeman, Bill Galloway, Ann Goddard, Bob Griffin, John Harland, Jay Hastings, Bill Heater, Allan Heater, Harry Heine, Konomu Kobo, Linda Nichol, Bill Proctor, Art Saunders, Fred Sharpe, Janet Stevenson, Peter Stoett, Diane Swanson, Gerry Truscott, whose editorial assistance was much appreciated, and Jim Wakelen.

Fred Sharpe and the Alaska Whale Foundation kindly provided images of humpback whales on the title pages for "The Quest for the Golden Egg" (p. 27) and "Requiem for a Humpback" (p. 59). Jim Darling, West Coast Whale Research, provided the humpback image on the title page for "Rose Harbour, 1911". Their assistance and support is much appreciated.

Bill Proctor is especially thanked for sharing his experiences with humpback whales.

Harry Heine's gracious offer of his water colour of a steam whaler chasing a whale for the cover of the book and his enthusiasm are gratefully acknowledged.

Finally, sincere thanks is due Jim Bisakowski and Lee Gabel of Desktop Publishing Ltd. Their patience, tact and good judgement have made this project a happy experience.

PREFACE

Perceptions of whales have changed since the last harpoon was fired in BC waters just 30 years ago. A great many people now think of whales more as friends than as fodder. How this transformation came about is a story in itself. It is not addressed here. This little book is intended simply to tell the story of how we interfaced with whales in an era when natural resources seemed limitless.

This is a discussion of commercial whaling in the Northeast Pacific Ocean from shore stations in British Columbia. Since the company founded in 1904 was bought in 1915 by an American, William Schupp, and expanded under the name of Consolidated Whaling Corporation to include a station on the coast of Washington and two in Alaska, whaling from those locations is mentioned, as is that of three other Alaska-based companies in the early part of the century. For the sake of brevity, modern shore-based whaling conducted along the coast of Oregon and northern California is not discussed.

Commercial whaling ended in BC before Greenpeace, based in Vancouver, launched its campaign against whaling by harassing pelagic whalers in the North Pacific. An intense campaign by volunteer organizations concerned with the environment, using the whale for a symbol, ensued. This campaign and the resultant 1986 moratorium on world-wide commercial whaling (exempting aboriginal whalers) by the International Whaling Commission is also not discussed here.

The commercial hunt on this coast began with a market for oil. Right whales, the source of marketable baleen, were rare. Fertilizer became an important product as modern shore whaling began, and later, whale meal. Whale meat did not become important until the last years of the whaling era, though there had been a short-lived market for salted whales' tails in Japan in 1906. Efforts to market whale meat during the First World War were largely unsuccessful owing to slow shipping, inadequate meat preservation and consumer prejudice. It was not until the Japanese firm, Taiyo Gyogyo, joined with BC Packers in 1961 to produce fresh-frozen whale meat for the Japanese market that meat became a significant product of the BC industry. Imported Japanese meat cutters working at the Coal Harbour station so relished whale meat that they would slice thin strips as they worked and sear them on the hot machinery for a mid-morning snack. Properly selected cuts, they claimed, were equal to the best beef.

While whale meat was the primary product of the BC whaling industry's final five years, it never caught on with Canadian consumers. Compared to insular Japan, where the land could not support the raising of livestock to feed a burgeoning population and whale meat had been consumed for centuries, Canada (and the US) had plenty of grazing space for beef. Professor Roy Chapman Andrews of the American Museum of Natural History in New York wrote, following a visit to study whaling in Japan in 1909, "In Japan it will be a national catastrophe when whaling ceases, because the diet of the ordinary native would consist of little besides rice, fish, and vegetables were it not for the thousands of tons of whale meat which are distributed fresh or canned to almost the entire Empire, and which furnish a healthful and palatable food at a low cost."

A full discussion seems warranted elsewhere of the continuing use of whales for food in a number of places around the world following long-established cultural practices.

The term "commercial whaling" as it is used here refers to the activities of newcomers to the BC coast when in actuality, the native peoples had carried on trade in whale products for centuries. Whales formed the basis of the economy for those tribes living on the rugged shore of Vancouver

Island and across the international border on the Olympic Peninsula. Containers of whale oil, an essential and much relished component of their diet, were used as a measure of wealth, given as bride price, and traded to inland tribes for items unavailable on the coast. Following the arrival of European settlers, the natives sold their oil for cash and trade goods in Victoria. Commercial though it was, native whaling never reached the excesses of the newcomers' hunt. Tempered by spiritual preparation, ritual proscriptions, and a respect for the whale; limited by the most basic hand-crafted implements; and dependent on extreme daring and endurance, perhaps it never could.

It is hoped that this brief review of BC whaling will afford the reader some background for understanding the developing debate around the future of whaling in the world's oceans.

ROSE HARBOUR, 1911

From Sealing to Whaling

Three blasts of a steamer whistle brought Captain George Heater to the deck of the sealing schooner *Jessie* as she rode at anchor in Rose Harbour one evening in late May, 1911. He fixed his gaze on Houston Stewart Channel, waiting for his brother's ship, the steam whaler *W. Grant*, to round Ellen Island with a tow of whales for the whaling station. Already, he had learned from the station manager, that William Heater had shot forty-seven whales in only a few weeks. That matched the record catch he had made the previous year in the station's first 3 weeks of operation. It was a remarkable record for the sealer-turned-whaler who had just learned the art of shooting whales and was already exceeding the catches of his Norwegian mentors.

William Heater was one of several sealing skippers who had joined the Pacific Whaling Company when the prospects for sealing in the North Pacific began to dim. Most of them were signed on as "pilots" of the whaling ships because of their knowledge of the BC coast. The Norwegian gunners who had delivered the boats from Norway stayed with them in BC as skippers, their experience indispensable to the new venture. Whaling activity in the North Atlantic had declined from over-exploitation, but Norwegian whalers were in demand in every ocean where their countryman Svend Foyn's modern whaling methods were coming into use. After a few years, most who had commanded the BC whaling ships would move on to Norwegian-owned operations in Alaska, Mexico and the Antarctic.

The Norwegian gunners were none too eager to teach Canadians how to shoot whales. They came from a tradition where "whale shooter" was a job with high status as well as high pay. Nevertheless, William Heater was coached by one of them and became the first BC whaler to get his own command: a whaling vessel brought from Norway in 1910 for the new whaling station in the Queen Charlotte Islands.

The *W. Grant*, towing three whales and listing slightly to port from the drag of an extra whale lashed to that side, soon appeared in the harbour entrance and eased toward the haul-out slip. A shout from the bridge assured the sealer that his younger brother had seen him. George Heater studied the little ship as her crew fussed about the float, tying up this trip's catch. He had seen her docked at Point Ellice in Victoria's upper harbour the previous winter, but this was the first time he had seen her at work. She looked much like the company's four older whalers and as good or better than the five new ones that had sailed around the Horn from Norway arriving in weather-beaten condition just before he had departed on this sealing voyage. It was hard to believe the *Grant* had been sent from Norway the previous year in sections aboard a Blue Funnel liner!

SS W. Grant, *Captain William Heater, entering Rose Harbour in the late 1930s.*

Sealing schooner Allie Alger, *Captain George Heater, 1904.*

Her whales secured, the 91-foot steamer moved across the tiny harbour and tied up next to the coal pile on the wharf to refuel while the captain of the schooner *Jessie* rowed over for a "gam" with his brother. They had not seen each other since the *Jessie* had left Victoria in early March. It would be late October back in Victoria when they met again. The schooner had come into Rose Harbour to leave 524 salted sealskins for tranship-

*Captain William Heater
posing at the gun of the
SS W. Grant, 1936 or 1937.*

ment to Victoria by way of the CPR Steamships' *Amur*. Now she was heading for the sea otter grounds and the Bering Sea.

When the refuelled *W. Grant* steamed back out to the Pacific a short time later, the two skippers' farewell waves were weighted with the recognition that the Heater brothers were no longer a pair as they had been in their 19 years of sealing in the north Pacific Ocean and Bering Sea, each commanding a schooner with native seal hunters and canoes aboard. They had come west together bringing their families from Harbour Grace, Newfoundland in the early 1890s, forced out of the maritime industry there by hard times. Both had done well. But now hard times in the Pacific sealing industry demanded another change of course.

Most of the sealers, like William Heater, had already abandoned sealing. Only five schooners had left Victoria in the spring of 1911 compared to between 60 and 70 in the early 1890s. The fur seals were disappearing. The controversy over conservation of the seal herds was being settled even as the two brothers talked in Rose Harbour: representatives of Japan, Russia, Great Britain and the United States had gathered in Washington, DC, to sign a treaty that would end pelagic sealing (offshore, from schooners). Only natives putting out from land in canoes and using traditional spears would be allowed to continue sealing. This was George Heater's last voyage after fur seals. His optimism and dogged determination notwithstanding, the catch would turn out to be meagre. The following year he would fit her out with a gasoline engine and take the *Jessie* halibut fishing.

William had taken a different tack. He was 45 years old in 1908 when he turned to whaling and, at age 78, was still in command of the *W. Grant* when the company ceased operating. He was so crippled with arthritis that his crew had to push him up the ladder to the bridge. Let anyone else shoot the whales? Never! He'd struggle down again and forward to the gun every time a whale was sighted from the lookout barrel on the mast. And that could be several times a day!

Steam Whaling in BC

The whaling company itself had been organized in 1904 by former fur sealers. Like the Heater brothers, Captain G. W. Sprott Balcom and his brother Reuben, master mariners from Nova Scotia, had been fur sealing in the Pacific since 1892. When it appeared that the industry was in trouble, they had returned to Halifax and organized the first pelagic fur-sealing expeditions to the South Atlantic and Antarctic. By 1903, the Balcoms had a fleet of six schooners making the 9-month voyages out of Halifax with handsome returns. But profitability waned in a few years. Using the corporate structure and financial backing of their South Seas Sealing Company, they returned to Victoria, and, changing the company name to Pacific Whaling Company, began catching and processing whales. Captain William Grant, managing director of the Victoria Sealing Company, who had invested with the Balcoms in South Atlantic sealing, was now a partner in whaling and became president of the company.

The Pacific Whaling Company opened its first station at Sechart in Barkley Sound on Vancouver Island's west coast late in 1905 with a modern steam-whaling ship, the *Orion*, ordered from Norway and delivered, under Reuben Balcom's command, by

Captain G. W. Sprott Balcom, founder of Pacific Whaling Company, posing at the gun of the SS St. Lawrence *in 1907.*

way of Cape Horn. Two other Vancouver Island stations were operating by 1907: one at Kyuquot and the other near Nanaimo at Page's Lagoon. The latter was closed down after only one season. Its catch of 96 humpback whales had apparently cleaned out the local population in the Straits of Georgia.

Over the next 3 years more, Norwegian-built catcher boats were acquired: three, the *St. Lawrence*, the *Germania* and the *Sebastian*, were second-hand; the fourth, the *W. Grant*, delivered from Norway in sections aboard the SS *Titan*.

In 1910, two more station sites were acquired in the Queen Charlotte Islands: Rose Harbour at the south end of the archipelago and Naden Harbour at the north end. Five new catcher boats were ordered from Norway, named *Black*, *White*, *Blue*, *Brown* and *Green*, expanding the whaling fleet to ten vessels plus a freighter, the SS *Gray*.

The new whalers, like the *Orion*, were sailed from Oslo (then Christiania) to Victoria by way of Cape Horn, a journey that took four months. Dr. J. N. Tønnessen with A. O. Johnsen in their definitive history of modern whaling, said of the *Orion*'s maiden voyage, "*It was a brilliant achievement sailing this small boat of 108 gross tons in mid-winter, through raging storms, across the North Sea and Atlantic, through the Roaring Forties and all the way up to Victoria. Whaling is often considered more the province of industry than of shipping, and for this reason, in the history of seafaring generally, little mention is made of the fact that some of the finest seamanship has been displayed*

SS Orion at the scene of the wreck of the Valencia, west coast of Vancouver Island, January 1906. Note that the aft mast has not yet been removed.

Crew of SS W. Grant, 1936. L-R, Astor Johnson, Tommy Hill, Oscar Ludvicson holding harpoon, Allan Heater, Louis Heglund.

*by crews on board tiny whale catch-
ers operating in the stormiest seas of
the world."*

Gradually all but a few of the
Norwegian crew were replaced
with local men; but the job of
shooting whales still remained
with the skilled men from Norway. It was an honoured calling,
even memorialized on tomb-
stones back home. The job
passed usually from father to son,
and few even of the Norwegian
crews were able to learn to shoot.

This presented a problem for
Sprott Balcom and the Pacific
Whaling Company. Sooner or
later Canadians should take over
the guns. From the start, Canadian skippers shared the bridge
with the Norwegians to satisfy
Canadian shipping regulations;
but they were paid half as much
as the master and were called
"pilots". Their local knowledge
of the BC coastline was essential
to safe and successful voyages.
For a few years, these seasoned
BC mariners watched the process
of shooting whales, until at last
one of the Norwegians was enlisted to train them at the gun.

William Heater was the first to take over his own command and the gun when the new *W. Grant* was delivered from Norway.

By 1911, each of the company's four operating stations had two or three whalers hunting within a radius of 30 to 50 miles, bringing in sometimes three to five whales at a time in the early bonanza years. Cruising at about 10 knots, the 91-foot ships were designed to turn quickly and could handle extremely heavy seas. There was bunker capacity for coal to last 6 or 7 days. Whalers inflated and flagged the whale carcasses, leaving them for collection later. The whale's tail was notched to identify which boat it belonged to. Drifting at night, they stayed out until they had enough whales to justify a tow to the station.

Among the eleven men who shared the damp, crowded, continually rolling confines of a steam whaler, there was a willingness to chase down the most elusive prey. Skipper and crew got a bonus for every whale caught according to its value in oil. In the early years, sperm whales paid the highest bonus because of their valuable case oil. Next in value were the much bigger sulphur-bottoms or blue whales, sometimes reaching 90 feet in length, and the fin whales. Least valuable, though most plentiful and closest to port were the humpbacks. Sei whales at first thought too small and low in oil to be of any value, would be hunted for their meat by 1917. Gray whales were not hunted; they were rarely even seen. Almost eliminated in the nineteenth century, they have only reappeared as a result of years of protection from uncontrolled whaling.

Whaling lasted from April until October. Though the ships could handle the west coast's violent winter storms, the tackle that brought in whales could not. Ships wintered at Captain Grant's wharf in Victoria's upper harbour. Each spring the captains, who were kept on retainer over the winter, supervised their outfitting and signed on their crew: a mate (whose role was formerly that of the pilot), four seamen, a cook, two engineers and two firemen. Once they stepped off Grant's wharf, they would rarely be ashore until the end of the season. Kora Larsen, seaman on the whalers

SS W. Grant *in Victoria's*
Inner Harbour, returning to
her winter berth.

in the 1920s thought the most exciting moment of the whole voyage was when the ship's whistle sounded for Victoria's Johnson Street bridge to raise, allowing the ship to return to her berth at Grant's wharf for the winter and the crew to go home. Occasionally a Chinese cook susceptible to seasickness would sign on quite unprepared for the physical duress of working in a rolling and pitching galley. It was said that the only reason he stayed aboard for the season was that there was no place to get off.

The most successful gunners were those with an intimate knowledge of whale habits. Spotted from the swaying barrel atop the mast, the whale's characteristic spout and movements indicated what kind of whale it was,

Steam whalers moored at Captain
Grant's wharf, Point Ellice, in
Victoria's Upper Harbour.
Whaling company's office visible
on the wharf; Captain Grant's
home, upper left.

whether it was feeding or travelling, and its direction and speed. Whales differed according to species in the length of time they would stay below, a critical factor in positioning the boat for a shot. Rising and dipping with the ocean swells as he manned the harpoon gun at the bow, the captain steadied himself with feet planted well apart, sighting down the cannon, waving signals to the mate on the bridge, waiting for his prey to surface. Not only must he be close enough when he fired the 90-pound harpoon; he must strike from the best angle to be successful.

Once the whale was hit and wounded, it had to be played in. Frequently a second explosive harpoon had to be shot to end its struggle, or a long hand-lance plunged deep. The process sometimes took hours. A whale might take off at such speed that even with the ship's engines in reverse they would be towed by the whale at 6 knots until it tired. Humpbacks had a habit of sounding and surfacing in unpredictable ways, often astern. The chase required cunning and patience.

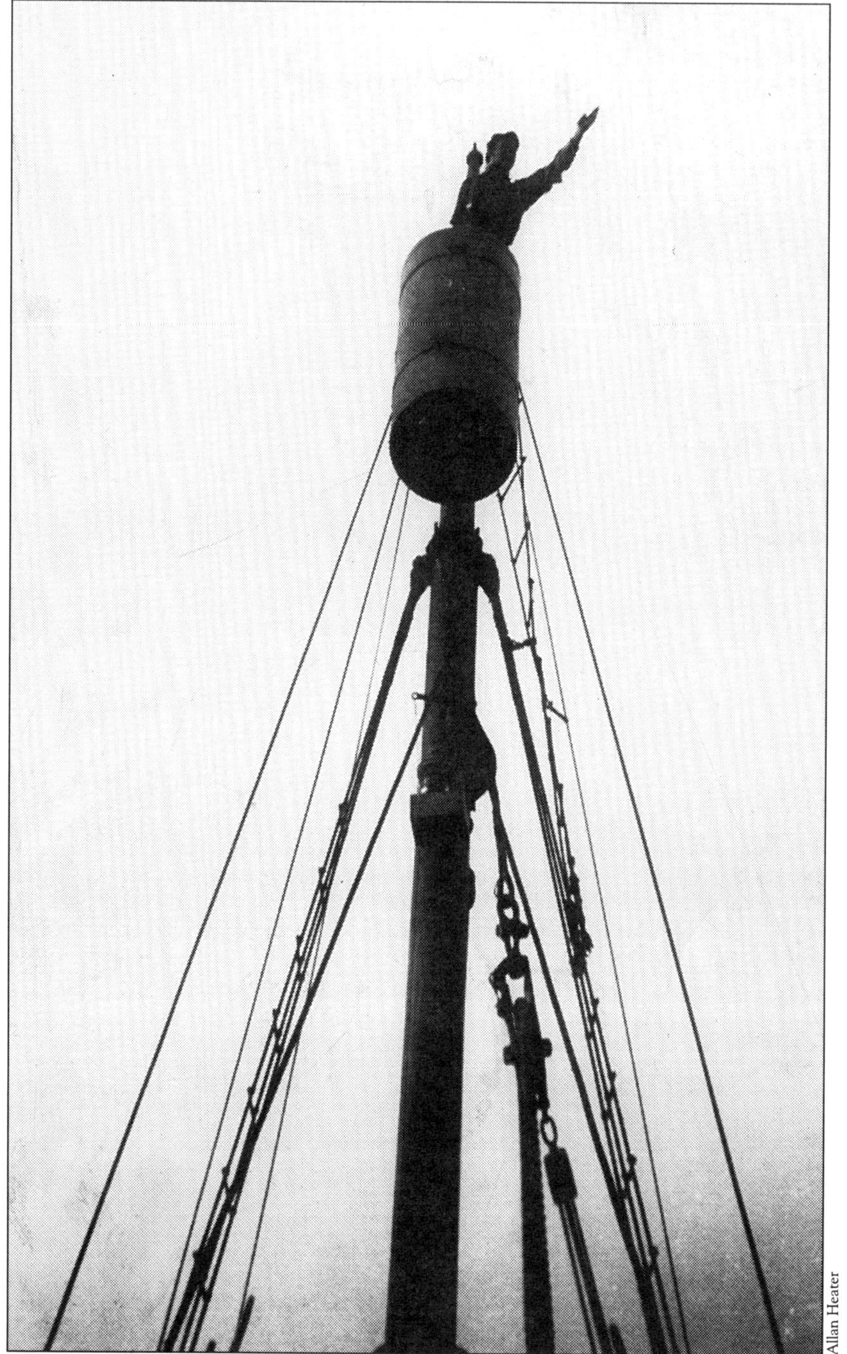

Whales off the port bow, according to the lookout in the barrel

Whaling Tales

Captain William Heater with a pot of tea on his fishing boat, Sarah. Whaling is ended for the season.

Captain Heater's whale-hunting technique was more than science. It was charged with a mixture of faith and determination that fixed his image in the memories of his crew for the rest of their lives. Heater singing "Jesus, Lover of My Soul" when his prey had sounded (dived); Heater swearing mightily when it surfaced behind him; Heater wearing his "sperm hat" with the little feathers on it for luck; Heater the last skipper to call it quits when the weather got too rough to whale. He had a theory shared by other whalers that sperm whales could be counted on to appear at the change of the moon.

In 1913, the Victoria *Times* reported that Captain Heater had steered his disabled ship back to Rose Harbour using a whale for a rudder. The harpooned whale, desperate to escape, had sounded so close to the stern of the boat that its tail had broken the rudder and two blades of the propeller. After eventually killing the whale, Heater had lines attached to its tail flukes and run forward "port and starboard". With the whale's nose tightly lashed under the stern, the "steering" lines were brought alongside the *Grant* to the huge winch drums on the forward deck. Heater, steering by pulls on the whale's tail, brought the ship slowly back to Rose Harbour. He even whale-steered her up to the wharf, according to the news story.

Captain Heater shooting a whale.

Whale-inflicted damage to the vessels was not uncommon, though in the history of BC whaling no ship was actually sunk by an enraged whale as happened to the *Lizzie Sorensen* in Alaska. Most damage was the result of whale carcasses hitting the side of the boat under tow. A report was made to the company headquarters in June 1916 that the *Orion*, whaling out of Rose Harbour, began leaking when an unexploded bomb protruding from the side of the whale they were towing pierced the side of the vessel. A temporary repair was made, and on arrival at the station, the vessel was listed to port while coaling so the blacksmith could make a permanent repair.

On the crescent-shaped stretch of shore west of the factory site at Rose Harbour, one can see at very low tide that a wide swath of the sea bottom has been cleared of boulders. This is where the whalers were careened for repairs. It was used in August 1934 to repair the *Green* which, like the *Grant* in 1913, had been struck by a harpooned whale. The whale was cut loose in order to save the ship. Canvass was stretched over the hole in the hull and crew bailed frantically for 6 hours. With the engine room flooded, she was helpless. Another whaler, the *White*, fortunately appeared on the scene and towed the *Green* back to Rose Harbour where she was beached, repaired, and sent whaling the following day.

A Norwegian gunner who stayed with the BC company years longer than the others was Captain Knut Halvorsen. A quiet bachelor, Halvorsen was

Loading the gun (turned backwards on its swivel). Black powder is being tamped in through the muzzle (forward end) to be followed with a wad of soft material. Next the harpoon (lying on the platform) will be loaded and whaling line attached. Finally a bomb will be screwed on and the gun turned forward. Two men are coiling the leading end of the whaling line (called foregoer) on the pan on the ship's bow.

Captain Knut Halvorsen, SS Brown,
waiting for calmer seas to shoot whales.

always welcome at the Rose Harbour station manager's house in 1916 to play a game of cards. It was a violation of company policy, for whalers were expected to be off hunting more whales the minute they had dropped their catch. But the manager, William Rolls, and his wife liked Halvorsen. It was a year of stress for Rolls. He was having difficulty with "something stronger than tea" somehow getting to the whaling crews. Little did he know that Captain Halvorsen and the crew of the *Brown* kept a still in Sperm Bay!

A Whaling Station

Standing today on the rust-encrusted rocks of Rose Harbour at low tide, with the fresh salty smell of seaweed in the morning air, a bald eagle's musical cry piercing the silence, one can barely imagine the stench and clangour of industry that identified the place in the whaling days. Other than a few relics of rusting machinery, little remains to indicate that this was once the site of a profitable whale-rendering plant and base for a fleet of steam whalers that brought in close to 4,000 whales during its 33-year lifespan.

Rose Harbour, scarcely more than half a mile across, is framed by Kunghit Island's low spruce-clad hills, with islands, rocks and shoals on either side. Facing the mountains of Moresby Island directly across the channel, it is completely sheltered from storms. Here, along its crescent shore, once stretched a line of buildings arranged like those of all the stations the company had operated.

Flensing a whale on the haul-up slip, Rose Harbour, 1930s.

Starting from the small creek at the south end of the beach and scanning northward, there were three bunk-houses and a cottage, then the haul-out slip with blubber loft behind where the whales' thick layer of insulating blubber was rendered into oil. Leading from the slip at an angle, were the carcass platform and sheds where the butchering and cooking took place,

Graeme Balcom courtesy Vancouver Maritime Museum

Drawing of Rose Harbour done for a company prospectus ca 1912.

then a smoke stack indicating the drier and sheds where cooked meat was pressed, dried, then ground into fertilizer (called, in the industry, "guano"). Finally there was the wharf, loaded with oak barrels ready to receive whale oil, and a pile of coal. A tall smoke stack behind the plant indicated the boiler house where a single man would shovel coal into six furnaces continually all day long to keep up steam power for the plant. A water tank stood on stilts nearby. The manager's house completed the scene.

In 1910, when the plant opened, the machinery was not new... not old either, having been used only one season in the ill-conceived whaling station at Page's Lagoon. It had been dismantled and shipped north to a much more promising site here at Rose Harbour. From the shortest-lived whaling station in the company's history to the longest.

The Shore Workers

George Huff managed the Rose Harbour station for just the first season. That may have been all he wanted. He'd had some unforeseen managerial challenges. His biggest crisis occurred just a couple of weeks before the whaling steamer arrived to begin work. At 9 o'clock on the evening of June 23, twenty-two of the Japanese workmen who had been building the station returned to their bunk-house to find Jentoro Kawasaki in a drunken rage. It was a scene not unfamiliar to Japanese labourers of that time. Isolated from family, bored, and with nowhere to spend their

Japanese flensers in the mouth of a humpback whale with station manager Harold Duckitt and engineer John Raine (in vest and shirt sleeves). Foreman Moichi Kosaka squatting. The baleen of a right whale can be seen behind the humpback.
Rose Harbour 1918.

Station hands cutting up blubber, Rose Harbour, 1936 or '37.

earnings, they entertained themselves with gambling. For a man who had lost everything at poker, home brew was like an explosive.

Kawasaki had been drinking for days. Brandishing an axe, he followed his countrymen upstairs, challenging them all to fight. Nobody, apparently, took him seriously except to complain that they couldn't sleep. When a senior member of the crew took the axe away from him, he went downstairs to the kitchen and returned with a butcher knife and a stick. He pulled Watakabe, a much smaller man, from his bunk. In the ensuing melee, bloody-and-bruised Watakabe ended up with the knife as they rolled on the floor. Kawasaki got up and staggered downstairs to the kitchen to rearm yet again. There he collapsed, lying on his face, another knife in his hand, with multiple slashes and two fatal stab wounds in his back. Kawasaki was carried to his bunk where he lay for three days wheezing air out of his wound. Medical care was out of reach.

On June 26, 1910, Jentoro Kawasaki died. George Huff, who had anxiously attended him, had his body laid in a rocky crevasse on Ellen Island, covered with branches, and sent Watakabe, escorted by three men, in a small boat some 25 miles to the provincial constable in Jedway. There, on July 16, the body which had been retrieved by the police launch from Ellen Island was viewed by a coroner's jury and a doctor. They concluded that Watakabe had killed Kawasaki in self defence. Oral tradition has it, however, that more than one hand, coming to Watakabe's defence, had plunged a knife in Kawasaki's back. Only the Japanese workers knew, and they never told.

Kawasaki's death inspired a "make work" program the following season to circumvent the drinking and gambling that inevitably accompanied idleness. To keep his men busy when fog or storms prevented the whalers from bringing in whales, Moichi Kosaka, Japanese foreman, devised a plan to build a "park". Trails were brushed out and pole benches, even a gazebo, were placed where there was a pleasant view of the channel.

Close to 100 men lived at Rose Harbour between April and October, most of them cutting up or processing whales. Japanese and Chinese made up the bulk of the work force, mainly because, as in the fishing industry, they could be employed cheaply. White labourers came on contract from the dying Newfoundland whaling industry. Each race was housed in a differ-

Guano shed, Rose Harbour, 1936 or '37.

Kinoe Kosaka on a "park bench" with her sons Isamu, standing left, and Gorge, seated on her left, and, left to right, station engineer's sons Arthur, Douglas and Nick Raine.
Rose Harbour 1916.

ent bunk-house, eating different food, speaking a different language. Homes , wives, children were an ocean or a continent away.

Moichi Kosaka was lucky. The foreman of the Japanese workers had his family with him. He had arrived in Victoria in 1905 with his wife, heading for Boston to study mining engineering. US immigration rules, however, which were more stringent than Canadian at the time, stopped him at the border because of an infected eye. So he took up residence in Canada. After five years of working in restaurants, Kosaka ultimately found a job supervising his less-educated countrymen in the whalery at Rose Harbour. His wife, Kinoe, cooked for the fifty-or-more men, and raised two small sons in their room in the Japanese bunk-house. Her motherly instincts extended to the many young men recently arrived from Japan to make a relative fortune for needy families back home. Kinoe Kosaka agonized over the pressures of loneliness and boredom that fueled the drunken gaming and fights that she heard upstairs, she told her grown children years later. Perhaps it was she who dreamed up the idea of building a park.

Each bunk-house had its own antidote for boredom. The Chinese, generally older men who had immigrated to Canada in an earlier era, were avid gamblers too. Solace for them was found in an opium pipe. Long after opium became illegal in BC, it was used in the Chinese bunk-house in Rose Harbour.

From the white men's bunk-house one might have heard an accordion playing a Newfoundland jig. In the early years of BC whaling, twenty or more men were brought west on 2-year contracts from Newfoundland, their skills no longer marketable in the colony. Whales had almost disappeared in the Atlantic and companies which had hastily organized to reap a bonanza were dying young.

Managerial staff had also been imported from Newfoundland for the fledgling Pacific Coast whaling industry. Their expertise was essential for the enterprise's success. Whaling station managers Scaplen, LeMarquand, Garcin, Ruck, Gosney, Rolls all brought their families across the continent to live in isolated whaling communities on the rugged BC coast.

With newspapers and mail brought once or twice a month, with no village in which to gossip and shop, with no church associations, with sometimes not so much as another woman to talk to, some of the women found the solitude hard to take. Visits from the passenger steamers bringing mail were big events. Those passengers whose curiosity was strong enough to overcome their revulsion to the stench of the place were greeted with delight as they came ashore to look around.

Rose Harbour had more families than the other stations during the first 8 to 10 years of operation. Children from three households, the station engineer's, the

Rose Harbour picnic, 1913. A reunion arranged by company superintendent S.C. Ruck (standing on log with Mrs. Ruck to his right). Nellie Gosney, centre, has been brought from Sechart on the SS Gray to visit her former station play-mate Dora Rolls, right. Rose Harbour manager William Rolls far left and his wife, Lena, seated. All Newfoundlanders.

manager's, and the Japanese foreman's, played along the rocky beaches, and hung out in the warm boiler house with the amiable fireman, Mike Benson, on rainy days. Sometimes they shot the blisters in heat-swollen whale carcasses with a .22-calibre rifle, atomizing the Rose Harbour atmosphere with essence of rotting whale. On one rare occasion, six children bounced on the tongue inside a huge whale's mouth. It was like a waterbed, one recalled 75 years later.

Racial animosity was high in the urban centres of BC at the time Rose Harbour was established. Yet the Kosaka family were simply friends to the manager's and engineer's families. The children were included in Sunday picnics and their mother, when time permitted, visited with the two other wives on the station. Being a teacher from a literary Japanese family and fluent in English, she prevailed upon the LeMarquand children's tutor, a maiden aunt, to include Isamu, her older boy, in their classes.

Newfoundland labourers, the expert flensers of the early years, were gradually replaced by Japanese and Chinese. These men worked 10 hours a day, 6 days a week for less than white men would have been paid. If there was a run of whales, they worked on into the night. After the whaling season was over, a crew of Japanese were retained at the station to do plant maintenance jobs.

Many of the whale processing jobs were extremely hazardous, and fatal accidents not uncommon. Men typically slipped into the vats of boiling oil. Without

Stirring a vat cooking meat and intestines for oil. Fatal accidents occurred here as workers fished scrap out of the boiling oil. Rose Harbour 1936 or '37.

radio or wireless, help could not be summoned. It sometimes took hours of waiting for a whaler to come into port and hours more for it to carry the injured man to the mainland hospital in Bella Bella where he died. One first-aid treatment was to smear the poor man's body with molasses and dust him with flour. The station first-aid man, doubling as book-keeper, had only the most basic supplies for medical emergencies.

The late Alan Armour, son of the last Rose Harbour manager, David Armour, and himself a former seaman for the company, erected a cairn in the summer of 1983 to the memory of those workers who had died in the service of the whaling company. Some of them, he believed, were still buried up in the woods behind the station.

An aerial view of Rose Harbour today shows a green and pleasant shore with a few homes and gardens. Yet closer examination shows that nature has outlined the old whaling station plan with trees. Rows and clusters of spruce grow where once stood factory and living quarters: wood giving life to wood. A string of spruce grows up the incline from the rotting wharf piles, nurtured by coal spilled from the coal car that ran between the wharf and the boiler house. At the top of a small meadow a miniature forest thrives in the ashes of the manager's house. And over the unused land a thick carpet of moss shrouds a dead industry's debris.

In the first decade of whaling on the BC coast, newspaper reporters covered the activities of the company, its fleet of ships and its stations with keen interest. Magazine articles described the exciting exploits of whalers and tried to explain the whale reduction process. None was successful in describing the most remarkable feature of a whalery: its smell. A Victoria newspaper reported in 1910 that longshoremen de-manded extra pay to unload sacks of whale fertilizer from the new Rose Harbour station. Their reason was that the smell would not come out of their clothes. The SS *Prince John* in later years stopped coming up to the wharf at Rose Harbour because the skipper claimed the putrid air actually

*Bottlenose whales were
so uncommon they called
for a photograph.
Kyuquot ca 1908.*

discoloured the ship's paint. "If you want your mail, you'll have to row out and get it!" he declared.

Mail comes even less readily now to Rose Harbour. It is dropped off by float planes if and when they have a charter hop that brings them to the south end of the Charlottes. Yet the few residents love the beauty and solitude of Rose Harbour and don't complain. Bone meal left over from the whale-processing days feeds their gardens; used brick and whale-oil-preserved lumber go into their building projects. Here and there relics of the whaling plant stand like weird statuary. Rotting wharf piles lean at crazy angles; rusting bone digesters loll where they fell through the crumbling carcass platform; brick furnaces pose as fortifications at the back of a vegetable garden; a steam winch lies askew like an agonized cadaver clenching a power no longer needed.

In 1862, Francis Poole, an English mining engineer, sailed to the Queen Charlotte Islands. As his schooner approached what 50 years later was to become the Rose Harbour whale-hunting grounds, he wrote: "Astern of us lies spread out the vast Pacific Ocean, completely alive with whales and porpoises......Who dare foretell how soon these frequenters of this half-known ocean path will be driven from the field of their sports, and their inheritance be taken possession of by the fleets of civilization?"

Not 2 years later, whales were being hunted with bomb lances in Saanich Inlet on Vancouver Island. The quest for the golden egg had begun.

Rusting bone digesters lying where they fell through the station floor 30 or 40 years earlier. Rose Harbour, 1988.

Abandoned steam winch. Rose Harbour, 1988.

THE QUEST
FOR THE GOLDEN EGG

An Overview of Commercial Whaling

Commercial whaling has been conducted for centuries with little thought for tomorrow. And what thought was given was based on poor understanding and flawed assumptions.

Only the native peoples who hunted whales for their own use and for trade maintained a sustainable hunt. Along the shores of Vancouver Island and northern Washington, the Nuu-chah-nulth and Makah de-

Natives prepare to cut up a whale at Echachis, near Tofino, 1909. One of the last traditional native whale hunts from Vancouver Island.

pended upon whales for nutritious oil and meat, and for baleen, bone and sinew, which had many uses. For thousands of years they utilized whales without diminishing their numbers. A balanced interaction existed between whale hunting peoples and their prey not only in the NE Pacific, but all around the world.

It was not until whale hunting became a business that over-hunting began. Whale products, particularly oil and baleen, became commodities with a price tag, and businessmen began financing voyages with an eye for a good return on their money. As commercial whaling grew, the distance between the source of capital and the resource expanded. Shareholders knew little about whales and their haunts and habits. Their only interest was a return on their investment. And the men who harvested the resource were dependent on their backing. The livelihood of both depended on the catch.

Makah whaler about to harpoon a whale, ca 1909. Asahel Curtis, photographer.

Washington State Historical Society

author's collection

Engraving showing artist's idea of how bowhead baleen was cut up in early commercial whaling period. "Fanons de Baleine, operations et outil du coupeur de Baleine," Benard Firexit.

Sail Whaling

Commercial whaling on the coast of British Columbia, which began less than 200 years ago, had its origins half-way around the world at a time when the Pacific Ocean was unknown to European mariners. From enterprising beginnings in the Bay of Biscay in the eleventh century, it grew over time to such proportions that the most valued whales became scarce.

When the whales became harder to find there were two alternatives for filling ships: going still farther afield or devising ways of taking the previously ignored species that remained. Technology served both strategies.

Larger sturdier ships were built with tryworks for boiling the blubber on deck–early factory ships–and steam afforded safe navigation in the ice. Guns that shot exploding lances led the way to a completely new era of mechanized whale hunting pioneered by the American whaling captain Thomas Welcome Roys and the Norwegian sealing captain and capitalist Svend Foyn.

As the hunt became ever more efficient, whales became harder to find. This did not deter the industry. It was taken for granted that there were always more whales over the horizon. Whale hunters had not been faced with the outer limits of growth; they lived in an era that extolled limitless opportunity for those who would venture forth.

Developing technology not only accelerated the harvest of whales; it increased the value of each whale caught. New whale products had to be developed as the market for whale oil was reduced by competing vegetable oils and petroleum, and the market for baleen succumbed to changes in ladies' fashions and the development of spring steel. To sustain profits in whaling, the Germans and Norwegians were processing whale carcasses into fertilizer by the late nineteenth century. In the twentieth century whales became animal food, canned meat and fresh-frozen whale steaks. The value of whale oil itself would revive when the new technology of hydrogenation was perfected in the early twentieth century, making it suitable for the manufacture of margarine, fine soaps and cosmetics. As profits from whales increased, the hunt intensified.

"Pursuit of the Greenland whale." Artist unknown. Wood engraving published 1850.

Pulling blubber aboard a whaling ship. Smoke rises from the tryworks where blubber is already being reduced to oil. "Dépècement d'une Baleine," ca 1850. Artist M. Bouquet. Line engraving by Rouarque.

When whaling for money first began, right whales close to the Atlantic shore were preferred because they were slow-swimming, rich in blubber and baleen, and did not sink when killed. They were the "right" whales to hunt. As these became harder to find, ships were built for whaling at sea. By the seventeenth century, the bowhead, similar to right whales in blubber and baleen, were chased in the Arctic waters around Spitzbergen and Greenland, and later west into Davis Strait and Baffin Bay. Eventually both right and bowhead whales on the Atlantic side of North America would all but disappear.

Venturing south in the Atlantic, whalers discovered the haunts of ocean-dwelling sperm whales. Their oil was already recognized as superior, for sperm whales had been utilized whenever they had washed ashore. Now whaling ships went after them at sea. At first blubber, peeled off the whale as it lay beside the ship, was carried back to port for rendering into oil; but efficiency demanded the invention of tryworks on board.

Technological innovation was combining with expanding markets to encourage the pursuit of whales to the ends of the earth.

Commercial Whaling Reaches the Pacific

Charles M. Scammon, *The Marine Mammals of the Northwestern Coast of North America*, 1874

Hunting gray whales in a lagoon on the coast of Mexico.

Commercial whaling in the Pacific Ocean began in 1789 when the first whaling ship rounded Cape Horn and discovered a bonanza. Soon whaling voyages from Atlantic ports would be extended to 3 years or more to harvest the new-found bounty. Over the next 60 years, hundreds of whale ships combed the Pacific. The sperm whales became harder to find. Much less valuable humpback whales became an alternative near the coast.

In 1846, still another alternative was discovered when two ships put into a lagoon along the Mexican coast and found the California gray whales' birthing place. Mothers and calves were easy targets in the Baja lagoons, which became a favourite haunt for whalers "between seasons".

The California gray whales' unique 12,000-20,000-km (7,500-12,500-mile) migration took them from Mexico to Siberia and back each year. For thousands of years, native hunters along the shores of Washington, BC, Alaska and Russia had hunted them, never depleting the whale stocks. But commercial hunting in the calving lagoons and then along the California coast would so reduce the resource for native people by the late nineteenth century that they gave up whaling. Siberian natives nearly starved.

Right whales had been discovered in the North Pacific in the 1830s, and the whaling fleet had quickly gravitated to what they called the Kodiak grounds between the BC coast and Aleutian Islands. In 1846, 292 ships

Hunting whales ca 1780. "Pêche de la Baleine." Artist, Morel; engraver, Rouarque. Line and stipple engraving.

sailed north of the 50th parallel encouraged by a strong demand for baleen. Within only 4 years, right whales were dwindling too.

Finally, north of Bering Sea, the last untouched whale resource was discovered: bowhead which were identical to those that had been hunted in the eastern Canadian Arctic. In 1848, Captain Thomas Welcome Roys of Sag Harbour, New York, ventured through the Bering Strait in a season of uncommonly fine weather and filled his ship with oil and baleen. He returned home the harbinger of a new era in northwest whaling. Within the next 3 years, 250 ships tried whaling in the Western Arctic, but many encountered ice conditions and weather that ruined their chances of a successful voyage. Nonetheless, the bowhead hunt would persist into the twentieth century assisted by new technology. Steam power was added to the whale ships, making them faster and more manoeuvrable in the ice.

The completion of the transcontinental railroad in the United States moved the centre of American whaling to San Francisco in the 1870s. Whale ships under sail and steam returned there with baleen and oil for the new west-coast refineries.

The Search for a New Technology

Whales had been killed for centuries with hand-held harpoons and lances from small rowboats. In the mid-1700s a flintlock gun was invented to fire harpoons from greater distances; it did not find much acceptance. But in the next century, the idea took off. Numerous new explosive devices came into use. Though the confrontation was intended to become safer for the whalers, in the experimental period many were injured. At the same time, death of the whale was less certain; many were mutilated and lost.

By the second half of the nineteenth century, all the sperm, right and bowhead whale populations in the world's oceans had been exploited. To keep the whaling industry alive, drastic innovations were called for. The solution, veteran Yankee whaler Thomas Welcome Roys believed, was to take whales that had been unreachable before: the fast-swimming rorquals. These, of which the blue whale was the largest—indeed the largest animal that has ever lived on earth—were not as rich in blubber, their baleen was shorter and much less marketable, and worst of all, they sank when they died. Nevertheless, the blue (also called sulphur-bottom), the fin, sei, Bryde's whale and eventually, the little minke, could provide a huge resource as yet untapped. The Norwegian capitalist and sealer, Svend Foyn, had come to the same conclusion half way around the world.

Nineteenth-century shore whaling on the California coast. The Greener's gun mounted on the bow of the whaleboat fired a harpoon. Once the whale was harpooned, the boat closed in and a bomb lance was fired into it.

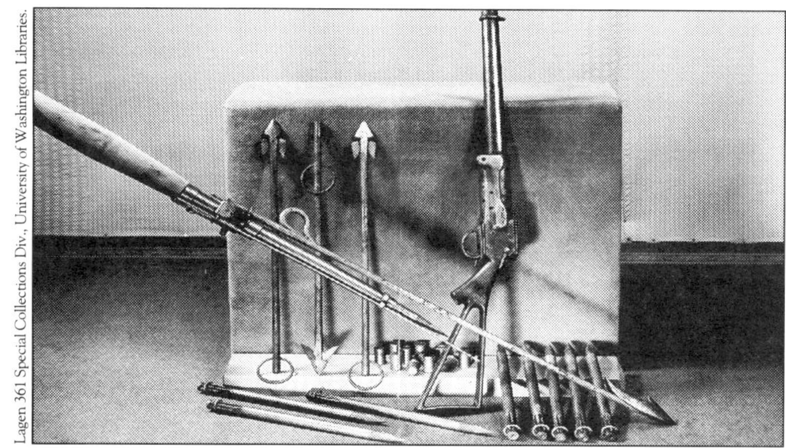

Nineteenth-century whaling guns. Darting gun (leaning diagonally) shows only the end of its 6-ft (2-m) handle. It fired a harpoon and a bomb-lance simultaneously. Shoulder gun leaning upright fired bomb lances (lying next to it) that exploded in the whale but did not have a line attached.

A new technology was required to hunt these fast-swimming whales. More had to be made from each whale killed and markets had to be found for the products. But there was still money to made in whaling.

Money for research and development, however, was not easily found in those days. Roys determined to find a way to hunt rorquals, but his resources were limited. He initially had financial backing, but it was withdrawn when results did not appear as soon as anticipated. Roys had to quit. Svend Foyn, on the other hand, had made himself a fortune. Determined to succeed, he was willing to spend his last krone on his own experiments.

Roys, with partner Gustavus Lilliendahl, went to Iceland in 1863, where they began to develop rocket-fired whaling harpoons and devices for winching up sunken whale carcasses. The latter was essential in the hunt for rorquals. The pair also pioneered a modern shore-whaling station to which they brought whales for processing. They worked for 5 years in Iceland before they ran out of funds and had to go home.

During their time in Iceland they were visited by Foyn, who had already in 1864 launched his revolutionary new steam-powered catcher boat, *Spes et Fides*, on whose bow he had mounted a cannon armed with a heavy grenade-tipped harpoon. After studying the Americans' rocket shoulder gun, Foyn was satisfied that his harpoon cannon mounted on a fast and manoeuvrable catcher boat was the best design.

Foyn also perfected a shore station where not only was blubber rendered, but every part of the whale was cooked, yielding an astonishing amount of additional oil. Finally, the dried meat and bone was pulverized into fertilizer. New technology had produced a new kind of whaling, later to be termed "modern shore whaling" that would spread around the world.

Commercial Whaling Begins in BC

Thomas Welcome Roys arrived on the coast of British Columbia in 1868, not long after his experimental foray in Iceland. He was one of several whalers who were attracted by reports of whales. Communication and trade had a north-south axis in the early days of the colony. Victoria news reached San Francisco twice monthly via the steamer *Pacific*. Ships plied, too, between Victoria and Honolulu, where the Pacific whaling fleet had converged for over 40 years. There were plenty of whalers in both ports who would prick up their ears at news of whales and carry their new-fangled whaling guns and traditional harpoons to Victoria.

Roys and others spent the next 6 years hunting whales, probably humpbacks, in the sheltered waters of the BC coast: Barkley Sound, Saanich Inlet, the Strait of Georgia, Burrard Inlet, Howe Sound and as far north as the Queen Charlotte Islands. Their bases of operation are still remembered in such place names as Whaletown, Blubber Bay and Whaling Station Bay. Shore stations were set up and the steamer

Schooner Kate, used for whaling in BC in 1860s by James Dawson and Captain Abel Douglass. Used for pelagic fur sealing from 1878 to 1898.

Commercial Whaling Sites in British Columbia

Nineteenth Century:

1. Mill Bay, Saanich Inlet
2. Whaletown, Cortes Island
3. Whaling Station Bay, Hornby Island
4. Pasley Island, Howe Sound
5. Jericho Beach, Burrard Inlet
6. Dodger's Cove, Barkley Sound
7. Knight Inlet
8. Cumshewa Inlet, Queen Charlotte Islands
9. Blubber Bay, Texada Island

Twentieth Century:

10. Sechart, Barkley Sound
11. Kyuquot (after 1918, Cachalot), Kyuquot Sound
12. Page's Lagoon
13. Rose Harbour, Queen Charlotte Islands
14. Naden Harbour, Queen Charlotte Islands
15. Coal Harbour, Quatsino Sound

Emma, the schooner *Kate* and the brig *Byzantium* were enlisted to carry whaleboats to the whaling grounds and blubber to shore. The schooner *Kate* was even lengthened by 15 feet to accommodate tryworks.

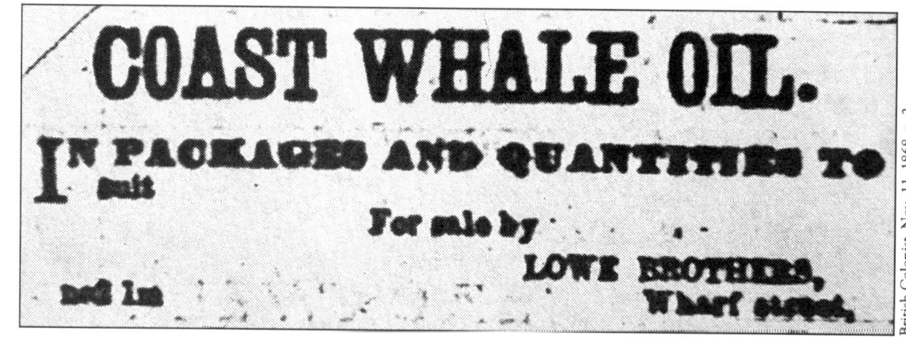

Newspaper advertisement offering BC whale oil.

These pioneering commercial whale hunts were considered big news and were reported almost weekly. Men blundered about with weapons that often did not work, wounding more whales than they brought to shore. Though the return in oil was small considering their effort, their optimism was reflected in the news accounts of their exploits.

Nineteenth-century whaling in BC waters waned after the colony joined the Canadian federation in 1871. The reason was clear: Canadian law forbade hunting whales with explosive devices. It had taken the whalers 6 or 7 years to reduce the stock of humpbacks in the sheltered waters of the Strait of Georgia. But during that time, Svend Foyn was perfecting a new whaling technology that would clean them out in 6 months when whaling resumed in 1907.

New Technology Changes the Hunt

Foyn's new technology revolutionized whaling. With increased efficiency, whale stocks were so depleted in north Norway that the country placed a temporary moratorium on whaling in 1904. The Norwegians then took their whaling show on the road, setting up whaling stations in nearby countries in the North Atlantic, and later in such remote places as South Africa, the Antarctic, Newfoundland, South America, and, finally, in Alaska. Their whale catchers, manned with Norwegian gunners and crews, were contracted for by countries all around the world, along with their whaling *matériel*.

When modern shore whaling reduced the whale stocks within the range of their catcher boats, a final great purge of whales was launched with the advent of factory ships, also pioneered by the Norwegians. Pelagic whaling, involving huge factory ships with fleets of fast and efficient catcher boats, could follow the whale herds as they migrated across the oceans. It would be pelagic whalers that made the final sweep of the North Pacific whales.

As more ships competed for whales, the return for each diminished, though the total catch rose with increased total effort. Those with the will to make whaling pay found a way. The trouble was that they would eventually reduce to commercial extinction the resource upon which the business of whaling depended.

Conservation Efforts

Newfoundland began Norwegian-style whaling in 1898, and by 1902 had enacted regulations intended to sustain the whale stocks. But even with amendments in 1904, the industry collapsed from over-hunting. It was, however, still an era of "try again somewhere else". Frustrated Newfoundlanders joined a growing crowd of Canadians clamouring for rights to whale off Canadian shores. Over fifty applications for Canadian whaling licences had been received from the east coast alone when, in 1903, Parliament cancelled the ban on whaling with explosive devices. In 1904, Canadian whaling regulations were enacted that seemed more protective of whale stocks, though in the end they did not sustain the whales. At least they insisted on complete utilization of the carcasses, and spaced whaling stations 100 miles apart, with only one catcher boat per station.

Norway led the way in conservation as well as in whaling technology. Her moratorium on whaling, passed in 1903, was the first of a number of moves to conserve the whale stocks. In 1927, having returned to whaling, Norway enacted regulations to forbid killing right whales, blue whales of less than 60 feet, and all species' suckling females and their calves.

International efforts to control whaling with a view to sustaining it began as early as 1913; but a formal agreement was not reached until 1931 when the Geneva Convention for the Regulation of Whaling was signed by eight countries. Though it failed to get the signatures of three active whaling nations, Japan, the USSR and Germany, it did initiate measures among some whaling nations to regulate whaling. In Canada and the United States, size limits were instituted and data collected about each whale caught. This information, to be used for scientific evaluation of the

INTERNATIONAL WHALING COMMISSION (IWC)

The world's whaling nations signed a treaty in 1946, the International Convention for the Regulation of Whaling (ICRW), which created the International Whaling Commission (IWC). The IWC's responsibility was the conservation of whale stocks and the orderly development of the whaling industry.

The IWC meets annually. Its membership, which now also includes nonwhaling nations, makes decisions to regulate hunting seasons, capture methods and catch quotas for each of the species of great whales. These decisions were to be based on data provided by the Scientific Committee of the IWC. In recent years the Commission has acted contrary to Scientific Committee findings, most notably in the continuation of a total moratorium on whaling. (The moratorium, a total ban on world whaling, was agreed upon in 1982 and went into effect in 1986 with the understanding that it would be reviewed by 1990 at the latest. Since then the moratorium has been renewed each year and remains in force at the time of this writing.)

The IWC does not have enforcement powers; member nations must police themselves.

Changes in whaling regulations must be approved by a vote of three-fourths of the IWC members casting votes.

Member nations that do not wish to comply with a new regulation may register an objection within a certain time frame and be exempt. Norway exercised this option with regard to the moratorium and conducts a managed hunt of minke whales for commercial purposes.

Member nations may also authorize whaling for research purposes (scientific whaling) without IWC approval. Both Norway and Japan conduct research hunts to permit science-based decisions in managing whale hunts and to improve scientific knowledge of ocean ecosystems. (Japan does not, at present, hunt whales commercially.)

Native whaling for food (subsistence whaling) was excepted by earlier treaties that attempted to protect over-hunted species such as the bowhead and gray whales. The 1946 treaty reaffirmed this exception; and in 1979 IWC regulations were expanded to recognize cultural needs as a valid basis for allowing continued whaling by aboriginal peoples, even on severely depleted whale stocks. Quotas for aboriginal whaling are set by the IWC for the United States and other member nations. Canada, which left the IWC in 1982, sets its own quotas for aboriginal whaling.

whale stocks, was forwarded to an International Bureau of Whaling Statistics in Norway. Right whales and suckling mothers and their calves were protected.

Meetings in 1937 to further regulate whaling were adopted by even fewer whaling nations. Signatories (including Canada and the United States, but not Japan and the Soviet Union) agreed to protect the gray whale as well as the right whale. The California gray whales had rarely been taken by 20th-century shore whalers on the Pacific Coast. They had been almost eliminated in the 19th century. They might have become prey to the industry as their stocks recovered had not their protection been assured in 1946. With protection, they had regained their pre-exploitation strength by the 1990s.

At last, the International Convention for the Regulation of Whaling was signed in 1946. It established the International Whaling Commission (IWC) whose mandate is to conserve whale stocks and maintain the orderly development of the whaling industry.

BC's Modern Shore Whaling Enterprise

Sechart whaling station, Vancouver Island, 1909.

Pacific Whaling Company, which began whaling in Barkley Sound in 1905 with Canada's first whaling license evolved through four corporate identities before it succumbed in 1946. Corporate expansion reached a dizzying pace by 1910 when the founders sold out to wealthy eastern railroad magnates, William Mackenzie and Donald Mann. Borrowing heavily, the new leadership, under the name of Canadian North Pacific Fisheries, added two new stations, doubled their whaling fleet to ten catchers and bought a freighter, the SS *Gray*. The next 4 years' catches from four stations–Sechart, Kyuquot, Rose Harbour and Naden Harbour–were: 1911, 1623 whales; 1912, 1109; 1913, 705; and 1914, 564. There would never be another year in BC whaling like 1911. It was all downhill for the overextended company. Bad whaling weather in two seasons and wartime marketing and transportation problems were given as excuses; but a more compelling reason for the company's bankruptcy was too few whales.

William Schupp, an American insurance broker with connections to Mackenzie and Mann, bought the company and renamed it Victoria

Kyuquot (Cachalot) whaling station, n.d. The second ramp was built in 1917 to handle whales to be processed for fresh or canned meat.

Naden Harbour whaling station, 1916.

Value of Whale Products Sold in 1918 by Consolidated Whaling Corporation

Oil	$880,823
Fertilizer (guano)	164,219
Whalebone (baleen)	18,368
Bone	9,598
Canned Meat	*14,180
Frozen Meat	**8,883

* balance of 1918 production sold in 1919 for $133,900 and 1920 for $8,416
** produced in 1917 and 1918

An aerial view of Rose Harbour whaling station in 1944.

Whaling Company in 1915. Three years later he combined it with two others he had acquired, American Pacific Whaling Company operating out of Grays Harbour, Washington, and North Pacific Sea Products operating out of Akutan, Alaska, to form Consolidated Whaling Corporation.

Between 1918 and 1943, Consolidated Whaling Corporation endured depressions, market slumps, labour crises, and ageing ships and machinery. Of the four Canadian whaling stations Schupp had bought in 1915, only the two in the Queen Charlotte Islands operated after 1925; and they flickered on and off like failing light bulbs. Naden Harbour processed its last whale in 1941, leaving Rose Harbour to go it alone. The American station in Grays Harbour closed in 1923. Alaska's Akutan station and another Schupp had built at Port Hobron were closed by 1939.

William Schupp, President of Consolidated Whaling Corporation, posing at the gun of the whaler SS Moran. n.d.

It was a blow to the company when the Japanese-Canadian flensing crew were interned with all west-coast residents of Japanese birth or ancestry in early 1942, leaving the work to inexperienced newcomers. Rose Harbour's last season was 1943. In 1946 the company was dissolved, and what ships had not been sold earlier were auctioned the following year.

Post-Second-World-War Whaling Resumption

Following the Second World War, interest in whaling revived, and a new consortium of fishing and logging interests, Western Whaling Corporation, began operations from a former seaplane base at Coal Harbour at the north end of Vancouver Island. Plagued, like its predecessor, with market and labour difficulties, it stopped whaling in 1959. Its major partner, BC Packers, resumed whaling just over a year later with a new partner, Taiyo Gyogyo, Japan's largest fishing and whaling company, under the name of Western Canada Whaling Company Ltd. While oil was still produced, emphasis was on preparing fresh-frozen whale meat for the Japanese market. The new improved whale hunt using large modern Japanese boats did not survive past 1967. Poor markets were given as the reason for the company's demise; but there were not enough whales left to make hunting worthwhile. Not only was the catch of 456 whales just 63% of the previous year's, but the whales were small and immature. Coal Harbour whaling came to a close, ending modern shore whaling in British Columbia.

Pacific Biological Station, Canada Department of Fisheries and Oceans, Nanaimo, BC

Fin whale at Coal Harbour whaling station. n.d.

Pacific Biological Station, Canada Department of Fisheries and Oceans, Nanaimo, BC

Carcass of fin whale after being stripped of meat (lower left). Coal Harbour, 1960s.

Pelagic Whaling Ends the Hunt

Japanese factory ship Nisshinmaru *used for research whaling in the western North Pacific. Length 130 m (390 ft).*

Pelagic whaling in the North Pacific had contributed to the demise of BC shore whaling. The use of floating factories in the Pacific had begun early. Both Norwegian whaling stations in Alaska had one in 1913. In the same year an expedition sent out from Norway to Magdalena Bay, Mexico, included a floating factory and three catcher boats. Though it was not a profitable venture, two similar Norwegian expeditions were sent out again between 1924 and 1928. These hunted whales off Mexico during the whales' breeding and calving season, with intervening hunts in the Bering Sea on their summer feeding grounds. In the 1920s and 30s American floating factories whaled in the Pacific with limited success.

Following the Second World War, Japanese and Russian pelagic whalers had moved into the North Pacific. It was they who conducted the final campaign in the war against the whales, competing on the eastern fringe of the Pacific with the whalers from Coal Harbour. In 1967 alone, they were reported to have taken 15,469 whales.

Commercial Whaling in the North Pacific Ocean and Bering Sea

1 1835-1850s: Right whales hunted over the Kodiak Grounds and in the Bering Sea. Bowhead whales also hunted in the Bering Sea.

2 1845-1870s: Gray whales hunted in their calving lagoons.

3 1854-1870s: Gray whales hunted from shore stations along the California coast.

4 1848-1908: Bowhead whales hunted in the Arctic Ocean.

5 1866-1872: Small whaleboats working from sailing vessels or shore stations take humpback whales in sheltered waters around Vancouver Island.

6 1905-1967: Modern shore-based whaling off Washington, Vancouver Island and the Queen Charlotte Islands take blue, fin, humpback, sei and sperm whales, and a few right whales and Baird's beaked whales. Protection of right, blue and humpback whales removes them from the catch before whaling ends. Gray whales are not hunted.

7 1907-1939: Modern shore-based whaling in Alaska from Admiralty and Baranoff islands off the SE coast, from Akutan in the Aleutians, and from Port Hobron on a small island near Kodiak Island. Blue, fin, humpback and sperm whales hunted, and until they are protected, a few right and gray whales.

8 1918-1970s: Modern shore-based whaling from northern California stations at Moss Landing, Trinidad, Fields Landing and Richmond, and at Warrenton, Oregon. Blue, fin, humpback and sei whales hunted, and before they are protected, a few gray whales.

9 1911-1946: Norwegian, Russian and American floating factories used at first close to shore stations for all commercial species of whales. Some California gray whales included in the early catches.

10 1954-1987: Russian and Japanese factory ships with fast diesel-powered catcher boats hunt sperm whales throughout the north Pacific Ocean.

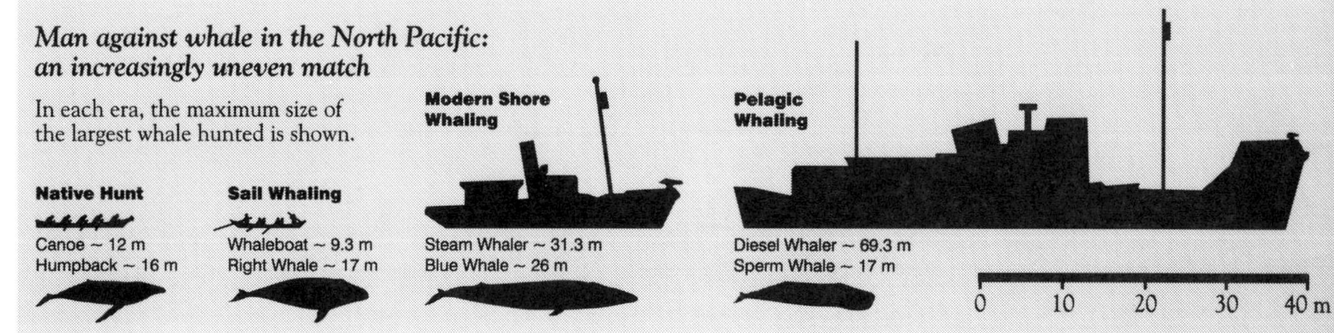

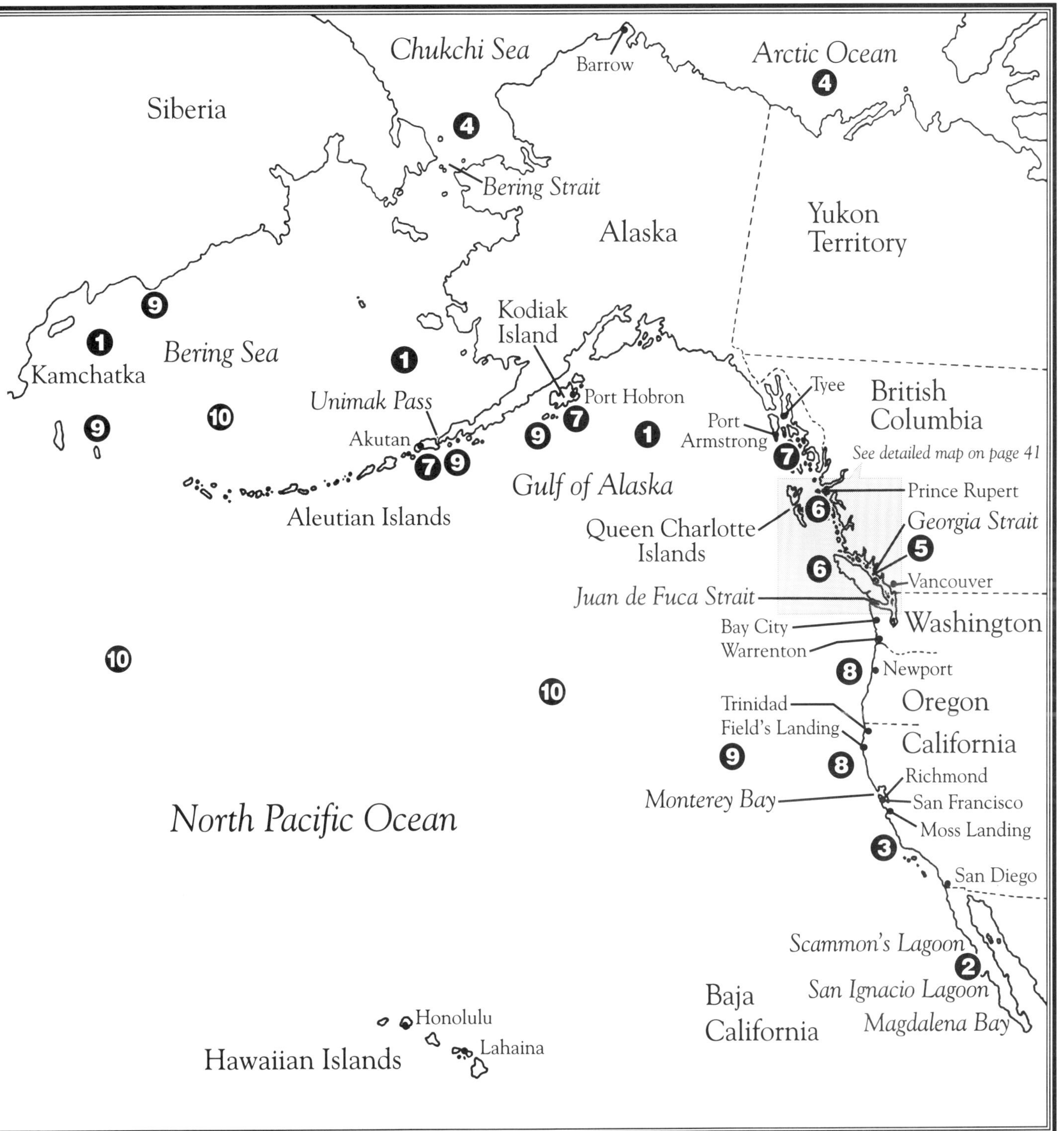

Chukchi Sea
Barrow
Arctic Ocean
4
Siberia
4
Bering Strait
Alaska
Yukon
Territory
Kodiak
Island
9
1
Bering Sea
1
Kamchatka
Port Hobron
Tyee
British
Columbia
Unimak Pass
7
Port
Armstrong
See detailed map on page 41
10
9
9
Akutan
1
7
7
Gulf of Alaska
6
Prince Rupert
Aleutian Islands
Queen Charlotte
Islands
Georgia Strait
5
6
Vancouver
Juan de Fuca Strait
Washington
Bay City
Warrenton
8
Newport
10
Oregon
10
Trinidad
Field's Landing
California
9
8
Richmond
North Pacific Ocean
San Francisco
Monterey Bay
Moss Landing
3
San Diego
Scammon's Lagoon
2
Baja
California
San Ignacio Lagoon
Magdalena Bay
Honolulu
Lahaina
Hawaiian Islands

Japanese catcher boat
Kyomaru *used for research*
whaling in the Western
North Pacific. Length 69 m
(207 ft).

BC's Position in World Whale Oil Production

The world's whaling industry was losing way as it competed with cheaper products in the oil and fat market. Redoubling their effort, Antarctic pelagic fleets consumed whales seemingly like a blue whale gulps krill. BC's share of whale oil production world-wide, though it had been 9% in 1909, became insignificant. During the 52 years of modern shore whaling in BC, a total of about 25,000 whales were taken. Compared to the Antarctic catch of 1,393,254 whales between 1904 and 1978, that may seem like nothing. Nonetheless, it was enough to bring an end to BC whaling.

The whaling industry in BC had been put out of business by the quest for the golden egg. Happily, though they came close, they did not kill the goose. Whale stocks world-wide may still return to healthy numbers under scientific management.

Requiem for a Humpback

Hunting Humpbacks

Humpback whales suffered greater depletion than any other whale species during the 52 years of BC's commercial whale hunt. Even though there was a bonus paid to gunner and crew according to the value of each species to the industry, with right whales (rarely found) paying the most, followed by sperm, sulphur-bottom (blue) and fin; and humpback paying the least. Humpbacks were most frequently taken. In the years 1908 to 1910, humpbacks comprised 70% to 90% of the catch.

Their habit of staying close to shore made humpbacks an easy target for the shore-based catcher boats; especially when bad weather threatened offshore. As long as humpbacks were accessible, gunners preferred to make more frequent short tows to the station with them than chance a longer cruise under uncertain conditions. Indeed, in 1906 humpbacks were so handy to the Sechart whaling station inside Barkley Sound, that the *Orion* hunted them during the first winter of operations while storms raged in the Pacific. Old whalers claim that they were "cleaned out" of the sound in that single season. The same thing happened out of Page's Lagoon on the Strait of Georgia. Their habit of feeding close to shore and following the herring runs into the sounds sealed the humpbacks' doom.

In the Queen Charlotte Islands, humpbacks, which had made up roughly half the catch when commercial whaling began in 1910, were reduced to about 5% by 1941. Protection enacted in 1965 came too late. Nevertheless, scientists believe the stocks are now recovering.

Losing a Friend

A query directed to residents of the sheltered waters on the "inside" of Vancouver Island regarding recent sightings of humpbacks brought the following communication from Bill Proctor of Simoom Sound.

This little story I wrote years ago about the humpbacks that lived here year round. I am 61 years old and have lived in the Knight Inlet area all my life.

I have been a commercial fisherman (troller) for 50 years.

As you will find when you read this story we were very fond of the old whales. They were around almost every day and they got to be just like one of the family.

This is about a humpback whale. Not just any old humpback but one I got to know real good.

When I was a small boy we lived in Freshwater Bay on the edge of Blackfish Sound. And this is where Old Joe the humpback lived. We would see him almost every day. We got to know his habits real good. He would come by on the tide and feed on the big schools of herring that were around there then. When he found a school he would dive and come up under them, and he would come out of the water head first. The herring would be running off his head. He would shake his head and he always looked like he was smiling.

He would feed like that every day at slack tide, and when the tide started to run hard he would go and lay in the back eddy by the kelp and sleep. He would lay there with his blowhole out and puff now and then. He seemed so content with his tummy full of herring.

Sometimes I would row up to him and sit and watch him for a long time. He knew I was there. But he didn't seem to care. And one time I looked under the water and his eye was open and he was looking back at me. I think he was smiling.

One day I let the rowboat drift till it was alongside of him, but he never moved and I reached out and touched him. He was so big and yet he seemed so gentle.

So it went on like that for a few years. Sometimes he would get in a playful mood and come in the bay and come under the floats. And he would go stand on his head in the kelp. He would have kelp hanging off his tail. But he seemed to enjoy it.

He was alone most of the time, but sometimes he had a buddy but not for long. But when they were together they would jump and roll around and slap the water with their flukes. They seemed to have a lot of fun.

There was a lot of humpbacks around back then and each one seemed to have his own place, and they stayed all year. They had good times for there was lots for them to eat and no one to bother them. So life was just a lot of fun for them.

One day a pod of killer whales went after him and he came in by the floats and the killers took off.

So Old Joe lived there for many years and I would watch him and it was uncanny the way he knew when the tide was going to slack.

He was sure a nice old friend.

There was a lady lived in Viner Sound and there was a humpback that lived in that area and he would go into Viner on the tide just about every day to feed. The people around there called him Barny because he had a big barnacle on his back. Sometimes he had a buddy so she was The Missus. So it was Barny and The Missus for the humpbacks of Fife Sound.

But this old lady that lived in Viner hated whales. So in 1951 she phoned the whalers out at Coal Harbour and told them there was some whales around Fife Sound and Knight Inlet. There was six lived up in Knight Inlet and they never left there. And it was the same in Kingcome.

Anyway the whalers came in and went up Knights and came out towing whales lashed to the sides of the boat. There was two whalers came in and one tow boat called the *Nahmint*. I will never forget that day in August. I was trolling in the mouth of Knight Inlet when I saw the old *Nahmint* coming out towing all the old whales alongside. I just about cried.

And that was the last we saw of the humpbacks for many years. Now we just see the odd one pass by.

I used to think a lot of Old Joe and Barny who had grown to trust humans, and then got shot by humans. It just never seemed right to me.

That is the way life is sometimes. But it was never the same without old Joe puffing along the kelp beds of Flower Island waiting for slack tide to fill his tummy.

Humpback whale feeding.

Jim Darling, West Coast Whale Research

EPILOGUE

Much has happened in the world of whales and whaling since Canada stopped commercial whaling in 1972. The California gray whale stock has returned to its pre-commercial-whaling size; other depleted stocks seem on the road to recovery. The smallest of the great whales, the minke, ignored by whalers earlier, has increased greatly, possibly, some scientists think, because it has less competition now for whale food.

Political, scientific, legal, even ethical concerns swirl in a deepening maelstrom of debate over whether to whale or not to whale.

And a limited hunt, controlled to sustain the whale stocks, goes on.

Appendices

WHALE SPECIES HUNTED IN THE
NORTH PACIFIC OCEAN AND BERING SEA[*]

Gray whale (*Eschrictius robustus*)
Historically called California gray whale, devilfish or mussel-digger. A baleen whale of maximum size 14 m (42 ft).

FOOD: Eats small crustaceans scooped from the sea bottom and plankton very close to shore.

RANGE: Migrates annually between mating and calving grounds in the bays of Mexico and the summer feeding grounds in the Bering, Chukchi and Beaufort seas. This annual trek of 12,000 to 20,000 km (7,500 to 12,500 miles) is the longest-known migration of any mammal on earth. Some grays spend the summer in sounds and bays along the route.

HABITAT: Stay close to shore on their passage to the Arctic in the spring, returning in the late fall somewhat farther out to sea.

HUNTED: For centuries natives of Siberia, the Aleutian Islands, the Gulf of Alaska, Vancouver Island and the northern coast of Washington State have hunted gray whales. Hunted almost to extinction in the mid-nineteenth century by sail whalers in the Mexican calving lagoons and

[*] Whale illustrations have been drawn proportionally to show their comparitive sizes.

shore-based whalers along the California coast, and periodically again after 1913 by pelagic whalers off Mexico, and Siberia.

Though not protected from hunting by Canada and the United States until 1937, rarely hunted from modern shore stations in the 20th century even as numbers increased, possibly because of their low yield of oil. Commercial whaling banned after 1946 by international agreement. Aboriginal hunting continues with quotas assigned by the International Whaling Commission. Population has returned at least to its size before nineteenth century commercial whaling.

Right whale (*Eubalaena glacialis*)

Also called black right whale or northern right whale to distinguish it from the southern right whale, *Eubalaena australis*, which is found in the southern hemisphere. Known in the north Atlantic historically as the Nordkaper and Biscayan right whale. A baleen whale of maximum size 17 m (51 ft).

FOOD: Eats small floating crustaceans (specifically copepods). Feeds by skimming the water surface with open mouth, filtering the feed through its long baleen plates.

RANGE: Inhabits temperate waters, migrating north to high-latitude feeding grounds in summer.

HABITAT: A slow swimmer that feeds close to shore.

HUNTED: Its high yield of oil from very thick blubber and its long baleen made it the first target of commercial whalers. Once abundant in all oceans of the world. Heavily hunted in the North Pacific between 1835 and the 1850s. Nearly extinct since the 1920s. Efforts at protection began in 1931. Protected from hunting by some nations, including Canada and the United States, as early as 1935. International agreement forbade hunting after 1946. Natives probably hunted the right whale. It is still rarely seen in the North Pacific.

Bowhead whale (*Balaena mysticetus*)

Also called the Greenland whale, Greenland right whale, Arctic right whale, great polar whale. A baleen whale of maximum size about 18 m (54 ft).

FOOD: Eats krill, and other plankton by skimming and filtering through long baleen plates.

RANGE: Found in the Bering, Chukchi and Beaufort seas. Migration follows the retreat and advance of the pack ice.

HABITAT: Lives at the edge of the pack ice.

HUNTED: Commercial hunting for bowhead, whose blubber and baleen were prized as much as the northern right whale's, reduced the population to near extinction by the early 20th century. As a right whale, it was protected from Canadian and American whaling as early as 1935. Universally protected from whaling since 1946, except when the meat and products would be used exclusively for local consumption by aborigines. Aboriginal whalers follow a quota. Though still very much depleted, the bowhead appears to be recovering.

Sperm whale (*Physeter macrocephalus*)

Also called the cachalot. Easily distinguished by its huge box-like head. A toothed whale of up to 17 m (51 ft).

FOOD: Feeds on squid, octopus and a variety of fish that live at great depths.

RANGE: Migration patterns bring groups of males into the northern limits of the Pacific while females and calves remain in warmer waters.

HABITAT: Inhabits the deep waters of the world's oceans, diving to depths of 3,600 ft and staying down for an hour or more.

HUNTED: Sperm whale oil and spermaceti was highly valued in the early nineteenth century. Markets created by the industrial age inspired a war on sperm whales in the southern part of the Pacific Ocean during the first half of the nineteenth century. In the North Pacific, however, sperm whales were not hunted until the twentieth century: first by modern shore whalers, later by intense pelagic whaling. Sperm whales were not taken by aboriginal whalers.

Baird's beaked whale (*Berardius bairdii*)

Called "bottlenose" by west-coast whalers. A toothed whale, it is the largest of the beaked whales, reaching to 13 m (39 ft).

FOOD: Feeds on squid, octopus, deep-sea fish and crustaceans.

RANGE: Seasonal movements are not known.

HABITAT: Found in deep temperate waters, it is native only to the North Pacific.

HUNTED: Caught infrequently by shore whalers in BC. Its oil was processed separately and sold as bottlenose oil.

Humpback whale (*Megaptera novaeangliae*)

A baleen whale of maximum size about 16 m (48 ft).

FOOD: Feeds primarily on krill and shoaling fish such as herring.

RANGE: Migrates from mating and calving areas in warm subtropical waters to northern colder waters for summer feeding. Some humpbacks appear to overwinter in sheltered BC waters.

HABITAT: Stays close to shore and follows spawning fish into bays and inlets. Whalers called it the beach comber.

HUNTED: Humpbacks were hunted by aboriginal whalers along the coast of Washington, Vancouver Island, the Queen Charlotte Islands and

Alaska since prehistoric times. In the early years of modern shore whaling they comprised most of the catch; but eventually they became harder to find. International agreement stopped the hunting of humpbacks in 1966.

Blue whale (*Balaenoptera musculus*)

Called sulphur-bottom by west-coast whalers until 1946. (The change may have been for uniformity in collating data collected for international whaling statistics) A baleen whale, the blue is the largest whale, and the largest animal ever to have existed. Very old blue whales, of which few are left, could reach 34 m (103 ft) in length. Most reach a maximum of 26 m (78 ft).

FOOD: Feeds mainly on krill (as much as 8 tons per day).

RANGE: Migration is seasonal from summer feeding grounds in northern latitudes to warm waters during winter.

HABITAT: Prefers deep water, but follows the continental shelf.

HUNTED: Because its great size afforded bountiful oil production, the blue whale was the primary target of modern whale catchers world-wide whose efficiency reduced the stocks to near extinction. In the North Pacific, blue whales became increasingly harder to find. In 1966 killing them was forbidden by international agreement.

Fin whale (*Balaenoptera physalus*)

Also called finback or finner by whalers. A baleen whale. In the North Pacific its maximum length is 24 m (72 ft). Its capacity for great speed (up to 20 knots) earned it the nickname "racehorse of the deep".

FOOD: Feeds mainly on krill though small crustaceans (copepods) and fish can make up as much as 30% of its diet.

RANGE: Migrates from winter breeding and calving grounds in temperate waters to feeding grounds as far north as the Bering Sea in summer, travelling thousands of miles each year.

HABITAT: Prefers deep water, but can also be found close to shore.

HUNTED: Apparently much more common in the North Pacific than blue whales, its proportion of the catch from BC shore stations increased as blue whales' numbers diminished. Young fin whales were valued for the fresh-frozen meat market in the last five years of Coal Harbour whaling. Victoria Whaling Company found a small market in Paris even in the midst of the First World War for fin whale baleen.

Fin whales were occasionally taken by aboriginal whalers.

Sei whale (*Balaenoptera borealis*)

A baleen whale that in the North Pacific can measure up to 16 m (48 ft). It looks like and swims as fast as the fin whale.

FOOD: Feeds on surface krill and copepods, small schooling fish and squid.

RANGE: Migration involves a general shift to northern waters in summer.

HABITAT: Lives well offshore in the North Pacific's temperate waters.

HUNTED: Though less rich in oil and relatively small, the sei was valued for fresh-frozen meat in the Coal Harbour whaling era in which the Japanese were partners. The large diesel catcher boats had access to their oceanic haunts.

 **Minke whale (*Balaenoptera acutorostrata*)**

Sometimes called the piked whale or sharp-headed finner. A baleen whale measuring up to 10 m (30 ft), it is the smallest baleen whale in North American waters.

FOOD: Feeds on shoaling fish such as herring and krill.

RANGE: From the Arctic seas to the equator. Migration patterns are largely unknown, though population tends to shift north in summer and south in winter.

HABITAT: Seen in coastal waters year-round, even penetrating loose ice in summer. They are called "bay whales" in Norway for their inshore haunts.

HUNTED: Minke have not been hunted commercially in the eastern North Pacific, and as a result, there is little data on their distribution or numbers. The Japanese are currently catching minke under the scientific whaling clause of the International Convention for the Regulation of Whaling to provide information on which to base a scientifically managed commercial hunt in the future. Minke are hunted commercially in Norway where their abundance is scientifically established.

WHALE PRODUCTS

OIL

Blubber, that thick white layer of fat under the whale's skin, was the first and always the chief source of whale oil. At first it was obtained by trying out (rendering) the blubber by applying heat. Tryworks were simply huge cooking pots over a fire. Steam coils inside the vats provided the heat for rendering oil in modern whaling operations, and lower grades of oil were cooked out of the whale's meat and bones. Whale oil could be used in its crude state for some purposes, and refined for others.

Oil from whales which in earlier times had great value for lighting and machine lubricants was replaced by petroleum and cheap vegetable oils after 1860. Although still sold for tanning, jute manufacture and industrial soaps, its commercial importance would not recover until the early twentieth century when hydrogenation was developed to turn it into fats suitable for margarine and fine soaps. During two world wars it provided glycerine for explosives.

Oil from sperm and other toothed whales differed so much from that of baleen whales that it was processed separately. It and the oil from spermaceti in the sperm whale's head were suited for the best lamp oil, for candles and for lubricating the finest instruments. In the nineteenth century it brought a much higher price and put a premium on catching sperm whales. But as markets changed it lost value. Soap makers, the main buyers of whale oil by 1925, could not use sperm oil. Developments in oil chemistry in the 1950s, however, brought renewed demand for sperm oil.

Whale species were rated according to the oil they could produce: the most valued coming from sperm whales; but the greatest quantity found in the fat bowhead of the Arctic, then right whales, blue, fin, humpback and sei in decreasing order. In the early part of the twentieth century, bonuses based on these values were paid to whaling crews.

Oak barrels had carried whale oil for centuries. Though steel drums and tanks in ships' holds would later carry oil, the barrel remained as the standard unit for reporting oil production. A barrel carried 31.5 US gallons of oil; six barrels of whale oil weighed a ton. Whales were even thought of in terms of barrels of oil. A "moderate-sized" sulphur-bottom (blue) whale was reported in 1907 to yield 40-60 barrels of the best quality oil from the blubber and 15-20 barrels of second quality oil from the carcass.

BALEEN (WHALEBONE)

Once a prized product from whales, baleen was also called whalebone, gillbone, finners and fins. Whalebone is definitely not the whale's bone, but thin flexible material, like fingernails. It hangs from the upper jaws of baleen whales, serving as a strainer to trap the millions of shrimp-like krill which make up their diet.

Right whales feed by scooping krill as they swim with open mouths. Their huge curved upper jaws have 230 to 300 long narrow plates of baleen hanging like the slats of modern vertical window blinds on each side of the mouth. The longest plates may reach 4 metres (12 feet), tapering from about 24 centimetres (1 foot) wide at the gum down to a point. Outer edges are smooth; but those next to the tongue are fringed, creating a sort of mat that traps the krill as the water leaves the mouth through the baleen plates. The longer the baleen, the more flexible it has to be in order to fold in when the mouth closes.

The other baleen whales have straight jaws and much shorter baleen ranging from just over a metre (3 feet) down to 16 centimetres (8 inches). The shorter baleen is more rigid. When these whales gulp huge quantities of feed their throats expand like big rubbery bags until the tongue squeezes the water out through the baleen. Long pleats or grooves reaching from the chin to the belly make this possible. These baleen whales are called rorquals from the Norwegian words for "grooved whale".

Commercial whalers, having started out hunting right whales and bowhead, found uses for whalebone as early as the 11th century. The fringes made tough durable brushes; the long flexible plates cut into thin strips found many uses later served by spring steel. Whalebone brought a high price in the nineteenth century when womens' fashions demanded hoops for skirts, stays and busks for tight-fitting corsets, and ribs for parasols and umbrellas. Limited uses for shorter whalebone persisted into the twentieth century.

Whalebone was removed from the mouth still attached to the gum, separated into individual slabs, scraped, washed, dried, split and cut into even lengths and bundled for shipment.

BONE

Until modern shore whaling developed techniques for processing the entire carcass of the whale, skeletal bone had little use. Jaw bones were used for arched gateways and because they were easily cut away, they could be sold to fertilizer companies before total utilization of the whale became common.

At the shore stations in BC, all the whale's bones were at first cooked along with meat and offal to extract more oil. By 1918 separate digesters for bones had been installed. With the oil cooked out of them, the bones were piled up for later grinding into bone meal which had a market as fertilizer.

WHALES' TEETH

Teeth from the sperm whale, the largest of the toothed whales, were not saleable in sperm-whaling days. They had great value to the crews of sailing ships, however, as a medium for carving during the long ship's passages home from the whaling grounds. The work, called scrimshaw, has value now as an art form.

There were no comparable spells of idleness on BC's shore stations in the 20th century; but sperm teeth were worked into cribbage boards, chess sets, knife and walking-cane handles, and even clock frames. A particularly talented Japanese flenser at Rose Harbour carved sperm teeth in the style of the sailing crews of an earlier era.

FERTILIZER (GUANO)

Modern whaling technology permitted the utilization of every bit of the whale. No longer were carcasses cast adrift in the sea; they were processed for extra oil, for bone for bone meal, and for fertilizer. Whale meal for animal feed and meat for human consumption would come later.

The manufacture of guano, as fertilizer was called in the whaling industry, was begun by the Germans in the late nineteenth century, adapted by Svend Foyn in Norway, and improved in North America by Ludwig Rissmüller, a German-American chemist. The patent for Rissmüller's "machinery for drying, grinding and screening meat and other substances" became the touchstone for the industry's success as the market for oil became more uncertain. Indeed the process was required in every whaling operation by Canadian law. Agriculture in California and Hawaii had plenty of use for whale guano.

Processing involved pressing the liquid out of the cooked carcass. What was left was then carried through a rotary drying drum that disintegrated and dried the material, finally screening it and delivering it with a blower to the hopper from which sacks were filled. The process created a cloud of guano dust that wafted over the station and settled on everything. Whaling station vegetable gardens flourished.

Guano was perhaps the chief offender in producing a whaling station's notorious smell.

WHALE MEAT

While salted whales' tails were shipped to Japan from as early as 1906, BC whaling stations did not make serious attempts at producing whale meat until 1917 when war-time meat shortages prevailed.

At first fresh meat was prepared for the market—63 tons in 1917 and 38 tons in 1918—but limitations of refrigeration and slow transportation produced an inferior product. A cannery built at Kyuquot produced 30,000 cases of whale meat (each with 48 1-pound cans) during 1917-1918. But the market for whale meat was lost after the war.

Grinding and drying the carcass residue into whale meal for animal feed rather than fertilizer was developed in the 1930s. That processed at Coal Harbour contained an average of 70% protein.

Some meat was sold for chicken and mink feed in the 1950s by Western Whaling Corporation—500 tons of fresh and frozen meat in 1953—as well as canned meat. But serious meat production would not begin until the Japanese firm Taiyo Gyogyo joined the BC company in 1962 to process fresh-frozen whale meat for the Japanese market.

Meat cutters were brought from Japan each season to select the meat from fin and sei whales, cut it up and prepare it for freezing and later shipment to Japan.

AMBERGRIS

Finding ambergris, a waxy substance occasionally formed in sperm whales' intestines, was every whale processor's dream. It rarely happened. Lumps as large as one thousand pounds have been recorded that saved a struggling whaling company from bankruptcy. A Norwegian floating factory's catcher in 1912 took a sperm whale off Tasmania with £23,000 worth of ambergris. In BC Captain Anderson was reported to have killed a sperm with $56,000 worth (which seems exaggerated) of ambergris which merited the gunner an extra bonus of $450. In 1922 Kyuquot workers found a 537-lb lump which reportedly was placed in a vault to be marketed in small amounts to keep the market price up. Ambergris was offered on the London market in 1934 by William Schupp for which the whaling company received $985 in US currency.

Finding it washed up on the beach is as unlikely as finding gold nuggets in a west-coast stream; but people still walk the sands with head down, looking for gray lumps of ambergris.

A CHRONOLOGY OF COMMERCIAL WHALING

Whaling in the NE Pacific with other important dates for reference.

11th century: Commercial use of whales begun in Bay of Biscay.

1500s: Basque fishermen cross the Atlantic to hunt right whales and bowhead from a shore station in Labrador.

1600s: British and Dutch hunt bowhead whales near Spitzbergen following Henry Hudson's report of whales seen in his attempt to sail to the North Pole. Whales at first rendered ashore. As ships range farther to sea to find whales, blubber stowed in barrels and taken to port for rendering.

1650: Colonists on Long Island, New York begin shore whaling. Nantucket residents begin whaling 20 years later.

1712: First sperm whale killed at sea off Nantucket marks beginning of American deep-water whaling.

1725: Vitus Bering sent to explore the North Pacific by Czar Peter the Great of Russia. Bering's naturalist, Georg Steller, notes many whales within sight of Mt. St. Elias, Alaska. Russians occupy Alaska, focussing on the sea-otter hunt, though in 1850, they will form the Russian-Finnish Whaling Co. to compete with American whalers. Crimean War limits whaling to only one successful season. Russians would later employ natives (Tlingit) in modern shore whaling out of Kilisnoo, Alaska.

1731:   Flintlock gun invented for use of British whalers in the Arctic is not well accepted. Even an improved version in 1792 is little used. Whaling guns will not be generally in use until mid-nineteenth century.

1750:   Right whale stocks on America's Atlantic shore are thinning out. Shipboard tryworks permit expansion of whaling. Sperm whale haunts found in Bahamas. Spermaceti candles invented, increasing sperm whale value. By 1775, 300 American ships have a monopoly on sperm whale fishery.

1774:   Spanish explorers reach the NW coast (Juan Perez at Queen Charlotte Islands 1774; Bodega y Quadra to SE Alaska 1775.)

1775:   American Revolutionary War keeps American fleet from whaling. British whaling ships fill the vacuum.

1778:   Capt. James Cook arrives at Nootka...observes native whaling equipment and whale oil trade. Other British explorers and traders begin to appear on the NW Pacific coast. John Meares' advocacy of commercial whaling and trade for whale oil from natives gets him in trouble with East India Company, which has a monopoly to trade on the Northwest Coast.

1789:   British whaling ship rounds Cape Horn and takes first sperm whale from Pacific. British whaling ships now permitted to travel through limited area in South Pacific. Though East India and South Sea companies have no whaling ships, whalers are forbidden to hunt whales in the Pacific outside of narrow corridor.

1790:   Nootka Convention settles dispute between Spanish and British over trade and shipping on the Northwest Coast. Though the East India Company would later stop objecting to British ships whaling in the North Pacific, none whaled. It was the French and Americans, free of political restraints, who would whale on the Northwest Coast.

1819:   American whaling ships begin using Lahaina and Honolulu as bases for refitting, recruiting, rest and lascivious recreation for hundreds of Yankee whalers absent from their New England ports for 3 to 5 years.

1820:   Japan Grounds become the most important whaling area of the North Pacific after sperm whales found there. Ships begin radiating out from Lahaina and Honolulu going to North Pacific in summer, reprovisioning in Hawaii, then to South Pacific in winter. By 1855, 500 ships will be whaling in the North Pacific.

1826:   Whalers begin anchoring in San Francisco Bay. By 1865 San Francisco will replace Hawaii as a provisioning port for whalers.

1835:   Capt. Barzillai Folger on the *Ganges* discovers right whales in what will become known as the Kodiak Grounds of the North Pacific, attracting 292 ships in 1846. Chief attraction to the right whales is whalebone (baleen), whose escalating value is dictated by ladies' fashion. By 1850 they will be severely depleted.

Late 1830s: Some sperm whalers spend the "between season" on the coast of Lower (Baja) California and in Monterey Bay hunting humpbacks.

1843:   Hudson's Bay Company moves its trading headquarters from Fort Vancouver on the lower Columbia River to Vancouver Island, anticipating the Oregon Treaty of 1846 which would set the boundary between the United States and British territory at the 49th parallel.

1843:   Two American whaling ships anchor in Shushartie Bay, north end of Vancouver Island, to spend the winter and do some trading with the Indians for furs. James Douglas, HBC administrator of the new Fort Victoria, sees them as a threat, saying, "If whales are found in the Sounds and Bays of the coast, the circumstance, when known, will attract so many vessels to these shores as to endanger the security of our trade." Winter whaling is not successful and the ships sail for Alaska.

1845:   Three whaling ships arrive in Victoria after unsuccessful hunt off Kamchatka, the right whales there having been fished out. Douglas reconsiders his policy of discouraging whalers when told many more would come the following year, and decides to trade necessities to them, but witholds items such as the best blankets that the Indians would trade for.

1845:   Calving lagoons of gray whales discovered by New England Captains James Smith and Josiah Stevens while bay whaling in Baja California. Exploitation of gray whales begins. Seven ships hunt gray whales in Magdalena Bay in 1846-47, and fifty in 1847-48. The hunt intensifies in 1855 after the introduction of bomb-lance guns that permit hunting the "devilfish" from a safer distance.

1846:   Boundary between US and Canada established by treaty. American whalers, unwelcome in Victoria, begin visiting Neah Bay, Washington and trading with the Makah for their whale oil. HBC officials who have a steady trade with the Makah also, try unsuccessfully to get them to refuse trade with the American whalers.

1847:   French whaling ship *General Teste* arrives in Victoria in September, with plans to whale in the Strait of Georgia until the following April. Whaling is unsuccessful and the ship leaves in February.

1848:   American whaler Captain Thomas Welcome Roys ventures through the Bering Strait into the Arctic Ocean after bowhead.

1849:   California gold rush attracts thousands of hopefuls. Whaling ship captains avoid San Francisco for fear of losing their crews. Disillusioned miners turn to shore whaling along the California coast using the new guns. Depletion of the gray whales is felt by natives along the entire migration route between Mexico and Siberia.

1859: Petroleum discovered in Pennsylvania. Whale oil, which had increased in value with the industrial revolution, is replaced with cheaper petroleum products. The value of whale and sperm oil for illumination fades with the development of kerosene (paraffin) from mineral oils, rape-seed (canola) oil, and ultimately, electricity. It had already been eclipsed in Britain by coal gas. By the late 1800s mineral oils (and vegetable oils) have taken over lubrication as well. Whale oil is still valued, however, in soap making, leather currying and in the manufacture of jute.

1863: Capt. Thomas Welcome Roys begins whaling off Iceland with partner Gustavus Lilliendahl using their new rocket-fired harpoons and taking whales to the first modern-type shore whaling station. They are visited by Norwegian Svend Foyn in 1866 who studies their innovations.

1864: Svend Foyn opens a modern shore whaling station in Finnmark, Norway, using a revolutionary steam-powered catcher boat, the *Spes et Fides*, which becomes the prototype for twentieth-century whale hunting. It carries an early version of the swivel harpoon cannon he had favoured over Roys' rocket gun. The station is the first to utilize all parts of the whale.

1866: Shore whaling begins on Vancouver Island with traditional harpoons and explosive whaling gear continues for 6 years. Entry of BC into Confederation of Canada would end hunting of marine mammals with explosive devices in 1871. Now restricted in their technology, most whalers turn to more promising pursuits such as fur sealing. One last effort at "traditional whaling" tried in mid-1890s at Blubber Bay, Texada Island.

1867: United States purchases Alaska from Russia.

1868: Pelagic hunting for fur seals begins with native hunters on schooner *Surprise*. Fur sealing would be a major maritime industry for BC into the first decade of the twentieth century. It would officially end by international treaty in 1911.

1869: Transcontinental railroad reaches San Francisco enabling that city to outfit ships. The US whaling fleet, greatly reduced during the Civil War, gradually moves to San Francisco, eliminating the need for 3-year voyages and the hazards of rounding Cape Horn. Recruiting in Hawaii is no longer necessary. Though the value of whale oil would drop by one-half in the next 10 years, that of baleen doubles, driven by the fashion industry. Ships still hunt for the now rare right whales in the North Pacific while they wait to get through the Bering Strait to the Arctic ice in search of bowhead.

1880: Use of steam auxiliary whalers for Arctic whaling begins. Captains head direct from San Francisco to Unimak Pass in the Aleutians, eager to reach the ice as soon as it breaks up. Their only reason to stop at Vancouver Island is for coal, for which steam whalers occasionally put in. Though whaling in the Arctic is apparently not of interest to the maritime community in Victoria, two steam whalers, the *Alexander* and the *Thistle* are sent north in the 1890s.

1898: First of several whaling companies establish shore stations in Newfoundland served by modern steam whalers built in Norway. A frenzy of whale hunting leads to decline in the whale stocks within 6 years idling shore crews and ships.

1903: Interest in whaling in Atlantic Canada prompts repeal of fisheries law that banned use of explosives in killing whales.

1904:   Shore whaling begins at the edge of the Antarctic on the island of South Georgia. Oil production is so successful that it depresses the world price of whale oil in 1905.

1904:   Canada amends the fisheries law to regulate whaling with modern harpoon cannon and sets regulations for modern shore whaling. From scores of applications on both coasts, one station begins operating in the Gulf of St. Lawrence in July and one on the Pacific coast in August of 1905.

1905:   Pacific Whaling Company, organized by Capt. Sprott Balcom, opens its first station on Vancouver Island at Sechart. Dr. Ludwig Rissmüller brought from Newfoundland as technical expert.

1907:   Two more Vancouver Island whaling stations built by Pacific Whaling Company: Kyuquot (renamed Cachalot in 1918) and Page's Lagoon.

1907:   American firm, Tyee Whaling Company, opens first shore-whaling station in Alaska at Tyee on Admiralty Island under Dr. Rissmüller's direction. It would survive until 1913. It's Seattle-built steam catcher *Tyee Jr.* deviates enough from Pacific Whaling Company's *Orion* to be considered peculiar by the Norwegians.

1909:   Whale oil becomes more marketable with the development of hydrogenation, which converts it to a solid and purified fat. It is now in demand for the production of fine soaps and cosmetics, and for margarine.

1910:   Pacific Whaling Company, merged with inactive Prince Rupert Whaling Co. and Queen Charlotte Whaling Co., is taken over by railroad magnates William MacKenzie and Donald Mann to form Canadian North Pacific Fisheries. Two more stations built in the Queen Charlotte Islands: Rose Harbour and Naden Harbour. Page's Lagoon, having whaled for only one season, is abandoned, its machinery moved to Rose Harbour.

1910: Capt. Sprott Balcom and Dr. Ludwig Rissmüller contract to build a whaling station at Bay City in Grays Harbor, Washington for newly organized American Pacific Whaling Company, a subsidiary of Canadian North Pacific Fisheries.

1912: Norwegian Peder Bogen, under the name of United States Whaling Company, builds a whaling station at Port Armstrong on Baranoff Island, Alaska. It is supplied with three steam catchers built in Seattle: the *Star I*, *Star II*, and *Star III*. It will operate until 1922. Some Norwegian gunners leave the BC company to work in Alaska with their countrymen. The floating factory *Sommerstadt* is sent to the Aleutians with two whalers, augmenting the oil produced on shore.

1912: Norwegian Lars Christensen establishes a second Norwegian company, the Alaska Whaling Company, based at Akutan in the Aleutian Islands. Two steam whalers *Kodiak* and *Unimak* built in Seattle and the Norwegian floating factory ship *Admiralen* is brought from the Antarctic. The company would reorganize as North Pacific Sea Products with American capital in 1914. Steam catcher *Tyee Jr.* is later bought from defunct Tyee Company and renamed *Tanginak*.

1914: Outbreak of the First World War affects Canadian whaling company's manpower, access to supplies from Norway, financing, distribution to overseas markets. Though price of whale oil goes up, it can't reach customary buyers in Glasgow. Embargo on shipping oil to US is eventually eased and US becomes main purchaser.

1915: Bankrupt Canadian North Pacific Fisheries and American Pacific Whaling are bought by American William Schupp. Canadian firm renamed Victoria Whaling Company.

1917: Sechart whaling station's last season processes only 90 whales compared to 474 in 1911. Station closed.

1917: War-time shortages of beef prompt effort to sell fresh and canned whale meat. A whale-meat cannery is built at Kyuquot. The product has poor appeal due to slow delivery of whales to the station, slow delivery of meat to the market, and inadequate refrigeration facilities. Consumer prejudice is enhanced.

1917: United States enters the First World War.

1918: William Schupp buys out North Pacific Sea Products, formerly Alaska Whaling Company. Machinery from Sechart sent to up-grade Akutan station.

1918: Schupp reorganizes American Pacific Whaling Company, North Pacific Sea Products and Victoria Whaling Company under one corporate umbrella, Consolidated Whaling Corporation. Headquarters are in Toronto, but American operations are directed from Seattle and Canadian from Victoria.

1918: End of the First World War. Stockpiled oil reaches the market in 1919.

1921: All Schupp's whaling stations idle due to international post-war recession and deflated price of whale oil.

1923: Schupp closes American Pacific Whaling Company's only station at Bay City, Washington because of poor catches of whales. The company will be dissolved in 1926 and its name transferred to the Alaska operations of North Pacific Sea Products.

1925: Kyuquot station (now called Cachalot, the French word for sperm whale) closed because of poor production.

1925: Norwegian Petter Sørlle's new design for a stern slipway on factory ships first put to use, opening a new era of intense oceanic whaling.

1926: Schupp opens a new Alaska station at Port Hobron, Sitkalidak Island, in the neighbourhood of Kodiak, using machinery from Kyuquot.

1929: World economic collapse.

1931: First attempt at international regulation of whaling, the Geneva Convention signed by twenty-six countries who agree to prohibit the killing of right whales (including bowhead) and female whales with calves and other conservation measures. Inoperative until ratified by at least eight governments (achieved in 1935). Three major whaling nations–Japan, Germany and the Soviet Union–fail to sign. Canada and US adopt its conservation practices and collecting of whaling statistics in 1935.

1931-32: Whale oil market again glutted by huge catches by pelagic whalers in the Antarctic, reducing the price of oil. Schupp again suspends all whaling operations.

1937: International Agreement to Regulate Whaling signed in London by only eight countries. Gray whales and right whales protected.

1937: Last season for Port Hobron whaling station in Alaska.

1939: Of Schupp's remaining three stations–Rose Harbour, Naden Harbour and Akutan–only Akutan in operation; this is its last season.

1939: Canada involved in Second World War, followed by the United States in 1941.

1941: Last season for Naden Harbour. Rose Harbour will survive two more seasons.

1942: Japanese residents of British Columbia interned. Rose Harbour forced to use inexperienced flensers to replace Japanese crew.

1943: Rose Harbour's last season. At close, whaling boats tied up for remainder of the war. American whalers taken over by US Government for Pacific coast patrol duty.

1945: End of Second World War.

1946: Consolidated Whaling Corporation declares bankruptcy and ships sold at auction the following year.

1946: All fourteen active commercial whaling nations, including Canada and US, sign International Convention for the Regulation of Whaling and become subject to the International Whaling Commission (IWC). Capture methods, hunting seasons and catch quotas of all species of the large whales will now be reviewed annually. Whaling statistics are to be forwarded to the International Bureau for Whaling Statistics in Norway. Protection of gray whales, right whales and bowhead from hunting now universal. Scientific studies to be carried out at Canadian and US whaling operations on the Pacific Coast.

1948: Fishing companies BC Packers and Nelson Brothers and Gibson Brothers logging interests form Western Whaling Corporation and begin whaling from Coal Harbour, a former seaplane base on Quatsino Sound at the north end of Vancouver Island.

1959: Poor market conditions and reduced production due to competition with Russian and Japanese factory ships. Closure of Coal Harbour station in 1960-61.

1962: BC Packers reorganizes in partnership with Japan's largest fishing conglomerate, Taiyo Gyogyo, forming Western Canada Whaling Company. Coal Harbour station reopened. Japanese whalers added to ageing fleet. Fresh-frozen whale meat for the Japanese market now the main product.

1965:   IWC protects sulphur-bottom (blue) whales and humpbacks from commercial hunting. A moratorium on all commercial whaling would not be in place until 1986.

1967:   Last season for Coal Harbour whaling station. Faced with poor markets and poor catch of whales, Western Canada Whaling Company closes, marking the end of commercial whaling on the BC coast.

1972-73: Canada and United States officially stop all commercial whaling.

GLOSSARY

Aboriginal (native): relating to whale and seal hunting, carried out by aboriginal inhabitants for their own nutritional and cultural needs.

Aborigine: an aboriginal inhabitant or native.

Ambergris: an amber-gray waxy substance secreted in sperm whales' intestines. Occasionally found while cutting up a sperm whale. Highly valued for making fine perfumes.

Bail: to scoop water out of a boat to keep it from sinking.

Baleen: fibrous plates in the mouths of whales without teeth. Used in trapping food. Also called whalebone, gillbone, fins and finners.

Barrel: also called "crow's nest". Referring to the look-out barrel on the mast from which a seaman watched for whales.

Barrel: unit of measurement for oil equal to 31.5 US gallons. Oak barrels of varying sizes, termed casks, were used to store and transport oil from early times. In the twentieth century these were gradually replaced with steel drums.

Bomb lance: brass tube containing explosives and a time fuse fired from a shoulder gun. No line was attached, as the whale was already harpooned.

Bottlenose whale: Baird's beaked whale.

Bow: the forward end of a ship or boat.

Beach: to run a ship onto a beach.

Berth: ship's place beside a wharf.

Blow (n): the visible mist in a whale's breath, varying in height and shape with each kind of whale.

Blow (v): referring to whales, to spout, expelling warm moist breath that condenses in cold air. "There she blows!", the now-romanticized announcement from the mast of sail whaling days evolved in modern whaling into several practical variations.

Blowhole: a whale's nostril through which it exhales (spouts) and inhales before diving. Toothed whales have one blowhole; baleen whales have two on the top of the head.

Blubber loft: upper floor of a whaling station building where blubber was rendered into oil.

Blubber: thick fatty layer under a whale's skin; source of whale oil.

Boiler house: building with furnaces and boilers that generated steam power.

Bone digesters: tanks in which whale bones were cooked to extract oil.

Bone meal: ground-up skeletal bone. Valuable for fertilizer.

Bonus: additional incentive pay for each whale caught, based on the whale's value. Paid to captain and part of the crew on whalers in BC, and to certain shore workers. Called "fish money" by some.

Bookkeeper: clerk at whaling station who kept track of workers' hours and pay earned, and who served as first-aid man.

Bottlenose whale: Baird's beaked whale.

Bridge: platform on a ship from which the officer in charge directs the ship's operation.

Bunk-house: building where workers slept and had their meals.

Cachalot: French word for sperm whale.

Calving lagoons: shallow stretches of salt water protected from the ocean by sand banks. Used by pregnant whales for giving birth (calving) and as a nursery for their young.

Cannon: a large heavy gun. On modern whalers, used to fire a harpoon tipped with a grenade (bomb).

Canoe (whaling): used by Nuu-chah-nulth and Makah people on the coast of BC and the State of Washington, a craft of 11 to 13 metres in length (33 to 39 ft) made from a hollowed cedar tree. Manned by eight men using sharply pointed wooden paddles.

Carcass: dead body. In whaling, often referred to the body stripped of its blubber.

Carcass platform: in shore whaling stations, a large area flanked by cooking vats where whales were cut up after their blubber had been removed.

Careened: referring to ships, allowed to lie over on one side in shallow water on a beach at low tide for repairs or painting. Incoming tide allows the ship to float and return to upright position.

Case: space in the forehead of the sperm whale outside of the skull which contains spermaceti.

Catcher boat (also chaser boat, whaler): Ship with a harpoon cannon mounted on the bow. Hunted whales and towed them to the shore station or to the factory ship for processing.

Cetacean: member of a classification of marine mammals (order Cetacea) that includes whales, porpoises and dolphins.

Coal car: small box-car on a railroad track that delivered coal from the whaling station wharf to a pile near the boiler house.

Coaling: loading coal aboard a steam ship; bunkering.

Commercial whaling: whaling as an income-producing industry.

Commercial extinction: level of whale numbers below which it is not profitable to hunt them.

Continental shelf: relatively shallow sea bed (to 200 m or 600 ft in BC) at the edge of a continent. Drops suddenly into much deeper water in some areas.

Copepods: small crustaceans floating in masses in the ocean.

Crude oil: in the whaling industry, oil that has been extracted from blubber and the carcass, but not further treated to remove suspended fine solid particles.

Crustaceans: animals of the class Crustacea which includes crabs, shrimp and lobsters. Those eaten by whales are very small, floating in the sea in large masses.

Diesel catcher boat: whaling ship powered by a diesel engine. Replaced steam-powered catcher boats in BC beginning in the 1950s.

Drier: a machine designed to dry, pulverize and screen material left after the whale meat and offal had been cooked to extract oil.

Ecosystem: a community of animals and plants and their environment.

Engineer: in steam whaling, person trained and certified to supervise steam engine use on whalers and at the whaling station.

Explosive harpoon: one with a device that explodes when it hits the whale. In modern whaling the explosive device was a heavy metal "nose" or bomb screwed to the leading end of the harpoon. The remains of the shattered bomb casing could be removed after the harpoon was recovered from the whale, and the harpoon straightened for continued use.

Extinction: reduction in numbers of individuals in a species to the degree that mating cannot occur and deaths therefor are not replaced by births.

Factory ship: ship with a stern slipway that permits hauling whales on board for processing. A self-contained factory free to operate on the high seas. Hunting is done by a fleet of catcher boats.

Feeding grounds: areas frequented by whales because of food available.

Fertilizer: also called guano. Made from residue of whale after oil was extracted.

Filling ship: 19th century term for getting a full cargo of whale oil before returning to port.

Fin: dorsal fin, a cartilaginous triangular extension on the back of some whales; absent in right whales and gray whales and poorly developed in sperm and humpbacks whales. Pectoral fins, also called flippers, are limbs evolved from front legs, shaped like paddles and adapted for balance and propulsion in whales.

Fireman: on a modern steam whaler and on the steam-powered shore station, man whose job was to keep up steam by shovelling coal into furnace.

Fish/fishing: commonly used for whale/whaling.

Flensing: Also flenching. Removing blubber from whales by scoring and cutting with long-handled curved knives. Strips then peeled off with power from a winch.

Flintlock gun: earliest gun (1731) designed to shoot a harpoon into a whale. Mounted on the bow of a whaleboat, it permitted harpooning at a much greater (and safer) distance than hand harpooning.

Flipper: see fin

Floating factory: ship modified for processing whale blubber that had been removed in the water alongside. Usually tied up in a bay. Later replaced by factory ships.

Flukes: two-lobed tail fin of whale, which propels by upward and downward motion.

Freighter: in BC whaling, also called tender ship. Used to carry coal and supplies to the stations and products back to the city. Workers were transported at the start and the end of each season.

Fur seal: *Callorhinus ursinus.* Northern fur seal hunted in the North Pacific for fur until 1911.

Galley: place on a ship where food is prepared.

Gam: a meeting at sea for conversation and exchange of news.

Grenade: a bomb containing a time fuse, screwed onto the forward end of a harpoon.

Guano: see fertilizer.

Gunner: one who shoots the harpoon cannon. In Norwegian the equivalent term is "whale shooter". The gunner was also the captain of the ship.

Harpoon: a steel shaft with hinged barbs to anchor it in the whale. Used for fastening line to a whale. Unless it had an explosive head, it was not expected to kill the whale. That was done with a lance.

Haul-out slip: sloping ramp used to pull whales from the water. Flensing was done on the slip.

Hull: the body of a ship.

Hydrogenation: a chemical process that solidifies fluid oils. Its application to whale oil turned it almost white and removed its unpleasant smell and taste.

Knot: unit of ship's speed: one nautical mile (1852 metres or 2025 yards) per hour.

Krill: small floating crustaceans, especially *Euphasia superba,* eaten by baleen whales. Dense concentrations of krill make the sea look red.

Krone: Norwegian unit of money.

Lance: long shaft with a pointed end used for killing a whale by hand after it has been harpooned.

Line: nautical term for rope with various uses aboard ship.

List: relating to a ship, to lean to one side.

Lookout barrel: tall barrel on the whaling ship's mast from which a seaman watches for whales.

Mast: on a ship, tall upright pole to support sails. Steam whalers delivered from Norway had two. Sails were used on the long passage to BC to save on coal and to stabilize the ships in heavy seas. The aft mast was removed soon after arrival; but the forward mast was required for the lookout barrel and for the easing gear that controlled tension on the whale line as a harpooned whale was being pulled in.

Mate: on a ship, the officer second-in-command.

Matériel: relating to whaling, specialized equipment such as harpoons, rope, etc.

Migration: in whales, the seasonal movements between areas suitable for bearing young and breeding, and those best for feeding.

Modern shore whaling: as opposed to "traditional" whaling, hunting whales with a powered catcher boat equipped with harpoon cannon and winches for pulling the whale to the ship. Whales towed to shore station or factory ship for processing.

Moratorium: a temporary suspension of an activity.

Native: see aboriginal.

Offal: intestines and other organs of an animal.

Opium pipe: a pipe used for smoking opium, an addictive substance imported from the Orient and, until 1908, legally manufactured for sale in Victoria.

Pelagic: of or performed on the open sea (as in seal hunting and whaling)

Pilot: a person authorized to take charge of a ship entering and leaving a harbour.

Plankton: masses of mainly microscopic organisms floating in water.

Pleats: grooves in the throat of certain baleen whales believed to allow their mouths to expand when gulping great quantities of food-laden water.

Port: place where ships come and go.

Port: directional indicator in navigation meaning "left" as you face forward.

Processing: manufacturing a whale into whale products.

Propeller (also, screw): . Curved blades on a shaft that turns, moving a ship through water or an aircraft through air.

Quota: a catch limit set to keep stocks from diminishing.

Recruiting: taking fresh provisions and supplies aboard ship.

Refined oil: oil that had suspended solids removed by filtering.

Rendering (also, trying out): Heating to turn fat to oil.

Retainer: pay during the nonwhaling season to assure continued service.

Rocket harpoon: a shoulder-fired explosive harpoon propelled by a rocket. Used in whaling as early as 1821 without attached whaling line. Roys' version had a line.

Rorqual: baleen whale with grooves or pleats on its throat allowing great expansion of the mouth.

Rudder: a flat piece hinged vertically to the stern of a ship for steering.

Sail whalers: sailing ships carrying whaleboats that pursued whales.

Scrimshaw: art of carving the teeth of sperm whales (and sometimes walrus), taken in the whale fishery, or using them or whale bone to make useful things.

Sea otter: a fur-bearing animal that lives in the sea close to shore. Their rich fur, about 1½ inches (3 cm) thick, was highly prized by natives and reserved for chiefs' wear. Once traders discovered their fur's value in China, commercial hunting of sea otters began.

Sealing schooner: schooner used in the North Pacific and South Atlantic for hunting migrating fur seals at sea. Men in small boats or canoes pursued swimming or sleeping seals with guns and spears.

Sealing: in the North Pacific, hunting fur seal on the open ocean for their fur. Not to be confused with the North Atlantic seal hunt that took hair seals on the ice primarily for their oil-rich fat.

Seaman: on a whaler, a crew member whose duties, in addition to standard ship-board tasks, included spending time in the barrel spotting whales, coiling whaling line (rope) in the line locker below decks, splicing whaling line, lancing and inflating whales, securing the whale's tail to the ship for towing, and later dropping it at the station float.

Shoaling fish: fish that swim in large multitudes or schools.

Shore whaling: whaling from a shore base to which whales were taken for processing.

Sinew: tough cord-like tissue that ties muscle to bone. Valued for its strength by native whalers, whale sinew also had a market in commercial whaling.

Skipper: master or captain of a ship.

Slip: a ramp up which whales were pulled from the water.

Sound: (n)an arm of the sea extending inland.

Sound: (v)to dive to the bottom.

Species: classification of animals or plants that have a common scientific name, common characteristics and are capable of interbreeding.

Spermaceti: white creamy substance in the case of a sperm whale's head. Valued for candles and cosmetics and a source of fine oil.

Spout: see blow.

Starboard: the right side of a ship as you face forward.

Steam auxiliary whaler: a sailing ship with auxiliary steam power. Sails were used on long passages to save coal.

Steam whaler: catcher boat powered by a steam engine, either coal- or oil-fired.

Steamer: ship powered by steam.

Stern slipway: ramp in a factory ship's stern that enabled whales to be taken on board for processing.

Stock: a group of individuals of the same species that inhabit a specific geographical area and rarely interbreed with those of the same species elsewhere.

Subsistence: harvesting, processing and consuming local resources using traditional knowledge.

Sulphur-bottom whale: name used historically for blue whale. Referred to colour of whale's underside, which in some individuals appears yellowish.

Sustainable hunt: a rate of hunting that keeps numbers of animals roughly in balance. Based on the theory that when animals are taken out, those left have more food, which increases the birth rate, thus replacing those taken.

Tack: to turn a vessel under sail to take advantage of the wind's direction.

Tow (v): to bring whales to the station or factory ship by tying them by the tail near the bow of the ship.

Tow (n): a trip to the station or factory ship with whales alongside.

Traditional whaling: sail whaling with whaleboats, hand-held harpoons and lances or small whaling guns, as opposed to modern powered catcher boats with harpoon cannons.

Trying out: see rendering.

Trypots: huge iron pots for rendering blubber into oil.

Tryworks: brick ovens with trypots set over a fire.

Vat: large metal tank for cooking.

Whale meal: meat dried and pulverized for animal feed.

Whalebone: baleen.

Whaler: in traditional whaling, the ship that carried whaleboats; in modern whaling, the catcher boat. A person who hunted whales.

Whalery: whaling station.

Whaling: hunting whales.

Whaling license: document issued by a government for a fee which authorizes certain activities subject to regulations. In Canada, whaling stations were licensed.

Whaling station: a self-contained community of workers around a whale-processing site. Whaling stations were occupied only during the whaling season, from April to October on the Pacific Coast, except for a small maintenance crew, a watchman and caretaker, or a manager and his family who had no incentive to spend winters in the city.

Winch: a mechanical devise to haul or raise great weights. Powered by steam on catcher boats and at the station in the days of steam whaling.

SOURCES

This brief history of BC whaling is drawn from company records and from personal interviews by the author with former participants in the whaling industry or their children, some of which are on audio tape in the BC Archives and Records Service.

Company records are housed in the Vancouver Maritime Museum, the Lagen Collection in the Manuscripts Division of the University of Washington in Seattle, the BC Archives and Records Service, and the private collection of Graeme Balcom. Whaling statistics are kept in the Pacific Biological Station in Nanaimo.

Some anecdotal material has been taken from the *Victoria Daily Times* and the *Victoria Colonist.*

An annotated discussion of this history may be found in "Reminiscences of Life of the Shore Whaling Stations of British Columbia 1905-1918" by Joan Goddard *Proceedings of the 17th Annual Conference on Underwater Archaeology,* Sacramento, California, 1986, Coyote Press, Salinas, California, pp. 206-211.

A 1992 statistical analysis by Linda Nichol and Kathy Heise entitled *The Historical Occurrence of Large Whales off the Queen Charlotte Islands,* prepared for the South Moresby/Gwaii Haanas National Park Reserve, Parks Canada was a valuable reference.

Whales, Dolphins and Porpoises of the Eastern North Pacific and Adjacent Arctic Waters, Stephen Leatherwood, Randall R. Reeves, William F. Perrin, and William E. Evans (Dover Publications, New York, 1988) was consulted for whale descriptions.

Discussion of 20th-century hunting of gray whales in the North Pacific was derived largely from Dale W. Rice and Allen A. Wolman, *The Life History and Ecology of the Gray Whale* (Eschrichtius robustus), Special Publication No. 3, The American Society of Mammalogists, published 30 August 1971.

SUGGESTED READING

Webb, Robert Lloyd. *On the Northwest. Commercial Whaling in the Pacific Northwest 1790-1967*. University of British Columbia Press, Vancouver, BC, 1988.

Hagelund, William A. *Whalers No More*. Harbour Publishing, Madiera Park, BC, 1987

Ellis, Richard. *Men and Whales*. Alfred A. Knopf, New York, 1991

Francis, Daniel. *A History of World Whaling*. Viking, Markham, Ontario, 1990

Spence, Bill. *Harpooned*. Crescent Books, London, 1980

Andrews, Roy Chapman. *Whale Hunting with Gun and Camera: A Naturalist's Account of the Modern Shore-Whaling Industry, of Whales and their Habits, and of Hunting Experiences in Various Parts of the World*. D. Appleton and Company, New York, 1928

Tønnessen, J.N. And A.O. Johnsen. *The History of Modern Whaling*. University of California Press, Berkeley, 1982

Scammon, Charles M. *The Marine Mammals of the Northwest Coast of North America*. Dover, London, 1968

Bockstoce, John R. *Whales, Ice and Men. The History of Whaling in the Western Arctic*. University of Washington Press, Seattle, 1986

Schmitt, Frederick P., Cornelis De Jong, and Frank H. Winter *Thomas Welcome Roys, America's Pioneer of Modern Whaling.* University Press of Virginia, Charlottesville, 1980

Merilees, Bill. "Humpbacks in Our Strait." *Waters.* (Vancouver Aquarium), 8 (1985)

Gordon, David G. And Alan Baldridge. *Gray Whales.* Monterey Bay Aquarium, Monterey, California, 1991

Hoyt, Erich, *The Whales of Canada.* Camden House Publishing, Cambden East, Ontario, 1988

INDEX